Anand Pandey, Ashish Goyal
Metal Cutting Processes

Also of interest

Metal Cutting Technologies.
Progress and Current Trends
J. Paulo Davim (Ed.), 2016
ISBN 978-3-11-044942-6, e-ISBN (PDF) 978-3-11-045174-0,
e-ISBN (EPUB) 978-3-11-044947-1

Metal Matrix Composites.
Materials, Manufacturing and Engineering
J. Paulo Davim (Ed.), 2014
ISBN 978-3-11-031541-7, e-ISBN (PDF) 978-3-11-048120-4,
e-ISBN (EPUB) 978-3-11-047871-6

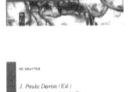

Drilling Technology.
Fundamentals and Recent Advances
J. Paulo Davim (Ed.), 2018
ISBN 978-3-11-047863-1, e-ISBN (PDF) 978-3-11-047871-6,
e-ISBN (EPUB) 978-3-11-067775-1

Maschinendynamik.
In Bildern und Beispielen
Marcus Schulz, 2017
ISBN 978-3-11-046579-2, e-ISBN (PDF) 978-3-11-046582-2,
e-ISBN (EPUB) 978-3-11-046597-6

Anand Pandey, Ashish Goyal

Metal Cutting Processes

—

DE GRUYTER

Authors
Dr. Anand Pandey
Department of Mechanical Engineering
Manipal University
Jaipur
Rajasthan
India
pandeyanand223@gmail.com

Dr. Ashish Goyal
Department of Mechanical Engineering
Manipal University
Jaipur
Rajasthan
India
ashish.goyal@jaipur.manipal.edu

ISBN 978-3-11-067656-3
e-ISBN (PDF) 978-3-11-067666-2
e-ISBN (EPUB) 978-3-11-067679-2

Library of Congress Control Number: 2021951790

Bibliographic information published by the Deutsche Nationalbibliothek
The Deutsche Nationalbibliothek lists this publication in the Deutsche Nationalbibliografie;
detailed bibliographic data are available on the Internet at http://dnb.dnb.de.

© 2022 Walter de Gruyter GmbH, Berlin/Boston
Cover image: Liuhsihsiang/E+/Getty Images
Typesetting: Integra Software Services Pvt. Ltd.
Printing and binding: CPI books GmbH, Leck

www.degruyter.com

MIX
Papier aus verantwor-
tungsvollen Quellen
FSC® C083411
www.fsc.org
FSC

Preface

The principle of metal cutting processes is utilized for shaping engineering materials shaping into real component/part manufacturing using machine tools and cutting tools. Accurate geometry of cutting tools plays a significant role in achieving machinability with superior surface quality and better dimensional accuracy. The book describes the conventional metal cutting processes such as turning, taper turning, milling, shaper, grinding, drilling, and other conventional machining problems and solutions with graphical representation. Each chapter is followed by several problems and questions that will help the reader significantly understand the formulas and calculations of machining responses.

https://doi.org/10.1515/9783110676662-202

Acknowledgments

We wish to thank our academic friends who have supported and motivated us to write the book in its simplest way so that all students can understand the real problems and their solutions related to machining of alloys and engineering materials.

We like to acknowledge and thank our students, faculty members, and technical staff who have assisted in writing the book in a new innovative teaching/learning, namely, outcome-based learning for solving problems of machining processes.

We sincerely thank the almighty GOD for giving us the wisdom to think, act, and transform our thoughts and experiences in the form of this book.

Authors

Dr. Anand Pandey

Dr. Ashish Goyal

https://doi.org/10.1515/9783110676662-203

Contents

Chapter 3
Metal cutting optimization —— 57

Chapter 4
Metal cutting performance responses —— **71**

About the book

The principle of metal cutting processes is utilized for shaping engineering materials into real component/part manufacturing using machine tools and cutting tools. Accurate geometry of cutting tools plays a significant role in achieving machinability with superior surface quality and better dimensional accuracy. The book describes conventional metal cutting processes such as turning, taper turning, drilling, milling, shaper, grinding, and other conventional machining problems and solutions in a graphical representation with conceptual outcome with change in parameter settings. The book illustrates problems/solutions that will help the reader significantly understand the formulas and the calculations of machining responses.

Salient features of book

- Strong emphasis on solving real metal cutting processes and problems with solutions.
- Calculation formulae
- Figures: 57
- Tables: 64

https://doi.org/10.1515/9783110676662-205

About the authors

Dr. Anand Pandey is presently serving as associate professor (senior scale) in the Department of Mechanical Engineering, Manipal University Jaipur, India. He has completed Ph.D. in machining of advanced superalloys using rotary EDM from Sant Longowal Institute of Engineering and Technology (Deemed University), Longowal, India, in 2013. Dr. Pandey is involved in writing books, editing book chapters, guiding the scholars for experimental analysis on advanced materials in the area of advanced machining processes (EDM, LBM), and reviewing journals/conferences. He has guided three research scholars for their doctoral degrees in the area of advanced machining processes.

Dr. Ashish Goyal is an associate professor in the Department of Mechanical Engineering, Manipal University Jaipur, India, since August 2012. He received his Ph.D. from Manipal University, Jaipur, for the research thesis study of machining of Inconel 625 superalloy by the wire EDM process. He received his master's degree from MNIT Jaipur in 2012 and B.Tech. from Rajasthan University in 2008. Dr. Goyal also served as a production engineer in the industry for one and half year. His area of research includes materials and manufacturing engineering, nonconventional machining process, and optimization techniques. Dr. Goyal has visited the University of Malaysia, Pahang, Malaysia, as a senior visiting lecture in June 2019 and do research activities and presented technical notes in international conferences. Dr. Goyal has published more than 25 research papers in reputed peer-reviewed journals and international conferences.

https://doi.org/10.1515/9783110676662-206

List of figures

https://doi.org/10.1515/9783110676662-207

List of tables

https://doi.org/10.1515/9783110676662-208

Chapter 1
Metal cutting processes

Learning problem 1

Lathe machine tool is used for manufacturing of cylindrical components using a single point cutting tool to shape the work specimen. Taper turning shafts are widely utilized in automobile commercial vehicles such as trucks, buses, cars for power transmission, and rotation of components. Mass production of tapered profile needs accuracy and dimensional precision with good surface finish. Determine the value of conicity of taper job having the length considering the type of skills develop in the graduate.

Bloom's taxonomy cognitive level ACTION VERB: determine

Expected outcome: knowledge (psychomotor domain) pertaining to skill development.

Big diameter = 100.5 mm and smaller diameter = 80.5 mm.

Learning solution 1

Formula of conicity

Length of taper, $I = 50$ m
Larger diameter, $D_1 = 100.5$ mm
Smaller diameter, $D_S = 80.5$ mm

$$K = \frac{D_t - D_S}{l}$$

$$= \frac{100.5 - 80.5}{50}$$

$$= \frac{20}{50} = 0.4 \text{ Ans.}$$

Now K versus length of taper.

Table 1: Values of length of taper and conicity.

L (in mm)	K (conicity)
50	0.4
100	0.2

https://doi.org/10.1515/9783110676662-001

Table 1 (continued)

L (in mm)	K (conicity)
200	0.1
300	0.066
500	0.04

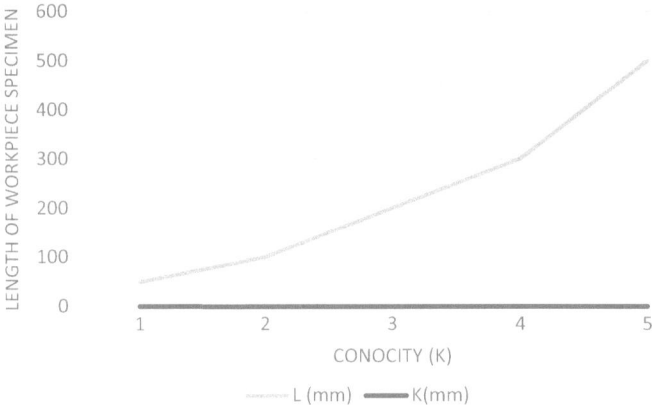

Figure 1: Graph between length of *w/p* and conicity.

The expected outcome for the graduate is to impart knowledge and computation skill for taper cutting and to understand that conicity reduces with length of the work material specimen.

Learning problem 2

Automotive components needed to manufacture in tapered profile with one side diameter to be larger than the other side for power transmission and motion of parts. Lathe machine tools, both conventional and CNC, are widely acceptable for taper turning of different engineering materials. Compute the larger diameter of a taper workpiece. The conicity for input values is to be computed, considering the length of taper cut to be 300.5 mm and small diameter as 85.5 mm.

Bloom's taxonomy cognitive level ACTION VERB: compute

Expected outcome: knowledge (psychomotor domain) pertaining to skill development.

Learning solution 2

Here, length of taper $= 300.5$ mm

Conicity, $K = 1/10$ to $1/50$

Smaller diameter, $D_S = 85.5$ mm

Large diameter $= D_e$

$$K = \frac{D_t - D_S}{l} \text{ (conicity)}$$

$$D_e = K_e + D_s$$

$$D_e = \frac{1}{20} \times 300.5 + 85.5$$

$$D_e = 100 \text{ mm}$$

Table 2: Values of large diameter and conicity.

K_e	D_e (mm)
1/10	115
1/20	100
1/30	95
1/40	92.5
1/50	91

The expected outcome for the graduate is to impart knowledge and computation skill for taper cutting process and to understand and remember and to apply the taper turning method using compound rest of the lathe machine tool for the experience that conicity reduces with smaller diameter of the work material specimen.

Learning problem 3

Metal cutting process, namely, taper turning compound rest is used to set the angle of cut for machining of hard materials (metallic) with machining single point cutting tool made of HSS (high-speed steel) or ceramic carbide inserts. Compute and apply the angle at which the compound rest is to be swiveled.

Bloom's taxonomy cognitive level ACTION VERB: compute

Expected outcome: knowledge (psychomotor domain) pertaining to skill development.

Tapered portion of the work = 100.5 mm

Smaller diameter (D_S) = 50.5 mm

Larger diameter (D_e) = 75.5 mm

Learning solution 3

Here, D_e = 75.5 mm

D_s = 50.5 mm

l = 100 mm

$$S = 1/100 \times \frac{900}{2}$$

= 4.5 mm

Dependency of S (angle) on conicity are given below.

Table 3: Values of conicity and angle.

K (conicity)	S (mm)
1/100	4.500
1/200	2.250
1/300	1.500
1/400	1.125
1/500	0.900

Table 4: Values of length and angle.

L (in mm)	S (mm)
700	3.5
800	4
900	4.5
1,000	5
1,100	5.5

The expected outcome for the graduate is to impart knowledge and computation skill for taper cutting process and to understand and remember and to apply the taper turning method.

Learning problem 4

Metal cutting standard practices on lathe machine tool process parameters, namely, feed rate and depth of cut play an important role in performance testing such as metal removal rate, lower tool wear, surface quality, and roughness. A turning tool has corner radius ranging from 0.35 to 1.55 mm. Determine and apply the feed rate in order to obtain a theoretical center time average roughness of 5.5 mm for Al-6061.

Bloom's taxonomy cognitive level ACTION VERB: determine and compute

Expected outcome: knowledge (psychomotor domain) pertaining to skill development.

Learning solution 4

In this case, $r_n = 1.5$ mm and $\text{hcLA} = 5.5 \times 10^{-3}$

$$\text{hcLA} = \frac{8+2}{10\sqrt{3r_n}}$$

$$f^2 = r_n \text{hcLA}/0.2566 = 0.02923$$

$$f = 0.17 \, \text{mm/rev}$$

Table 5: Values of corner radius and feed rate.

r_n (in mm)	f (mm/rev)
0.3	0.076
0.6	0.108
0.9	0.132
1.2	0.152
1.5	0.17

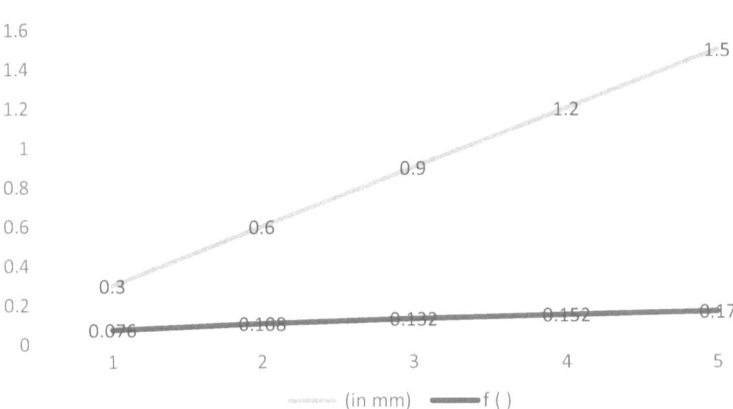

Figure 2: Graph between feed rate and center time average roughness.

The expected outcome for the graduate is to impart knowledge and computation skill for taper cutting process and to understand and apply the surface roughness measurement/testing.

Learning problem 5

Machinability is referred to as ease of machining and depends on machining factors, namely, RPM of chuck, feed rate, depth of cut, and type of material (hardness). Calculate and determine the machining time required for mild steel rod to reduce from 60.5 to 50.5 mm diameter for a length of 1,500 mm and depth of cut 2.5 mm for rough cut and 1.5 mm for finish cut.

Learning solution 5

$$\therefore T_m = \frac{A+l+O}{fr \times N} \times n (\text{machining time calculation}) \,\&\, V = \frac{\pi DN}{1,000} = \frac{\pi \times 60.5 \times N}{1,000} = 30 \text{m/min}$$

$$\text{or } N = \frac{30 \times 1,000}{\pi \times 60} = \frac{500}{\pi} \text{ RPM}$$

$$T_m = \frac{5 + 1,500 + 5}{0.5 \times \frac{500}{\pi}} \times 3 \text{ min} = 56.93 \text{ min}$$

Depends on T_m and cutting speed (V).

Table 6: Values of cutting speed and T_m.

Cutting speed (V)	T_m (time taken for machining in min)
30	56.93
40	42.69
50	34.15
60	28.46
70	24.39

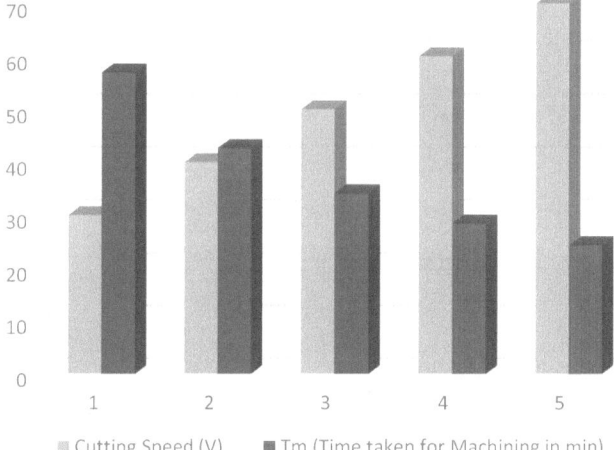

Figure 3: Graph between cutting speed and T_m.

The expected outcome for the graduate is to impart knowledge and computation skill for calculation of machining time as cutting speed increases.

Learning problem 6

Manufacturing of geometrical feature (thread cutting) using lathe machine tool is widely applicable in small-scale industries to large-scale industries for fastening and power transmission industries (automobile parts are required for fastening screws, nuts, and bolts). Calculate and determine the time of thread cutting required for external thread cutting of 2.5 mm. Pitch thread on job of 20.5 mm diameter, namely, the work material: die steel rod at a cutting speed of 8.5 m/min for a length of 100 mm on a shaft of 200 mm.

Bloom's taxonomy cognitive level ACTION VERB: calculate

Expected outcome: knowledge (psychomotor domain) pertaining to skill development.

Learning solution 6

Machining time for thread cutting

Diameter of the job,	$D = 20.5$ mm	
Length of threading portion,	$l = 100$ mm	
No. of cuts for external thread	$= 2.5 \times$ pitch	
	$= 2.5 \times 2$	
	$= 5$	

$$\therefore T_m = \frac{l \times \text{no. of cuts}}{N \times \text{lead}}$$

$$\text{and } N = \frac{V \times 1,000}{\pi\, D} = \frac{8 \times 1,000}{\pi \times 20.5}\, \text{rpm} = \frac{400}{\pi}\, \text{rpm}$$

$$T_m = \frac{100}{\frac{400}{\pi} \times 2} \times 5 \text{ min} = 1.96 \text{ min}$$

Depends on RPM and T_m.

Table 7: Values of RPM and diameter of job.

RPM	Diameter (D)
$\left(\frac{400}{\pi}\right) 127.32 \approx 127$	2.92
$\left(\frac{400}{\pi}\right) 159$	2.35
$\left(\frac{400}{\pi}\right) 191$	1.98
$\left(\frac{400}{\pi}\right) 223$	1.62
$\left(\frac{400}{\pi}\right) 254$	1.49

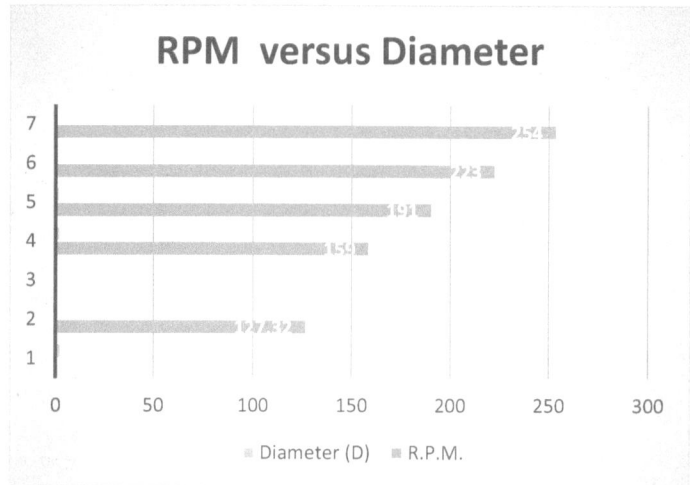

Figure 4: Graph between RPM and diameter of job.

The expected outcome for the graduate is to impart knowledge and computation skill thread cutting practices on cylindrical profile components.

Learning problem 7

Gear manufacturing from macro- to micro-size is required almost in all fabrication and power transmission industries globally. The series of gears that drive the lead screw are called change gears as they may change them to turn different thread pitches. To understand it is required to calculate and determine the change gears to cut a single assuming start thread as 1.55 mm pitch value lead screw of 6.5 mm pitch for aluminium-6062 material.

Bloom's taxonomy cognitive level ACTION VERB: calculate and determine

Expected outcome: knowledge (psychomotor domain) pertaining to skill development.

Learning solution 7

$$\text{The change in gear ratio} = \frac{\text{Pitch of the 30b} \times \text{no. of starts}}{\text{Pitch of the lead screw}}$$

$$= \frac{\text{Driver gear(S)}}{\text{Driven gear(S)}}$$

$$= \frac{1.55 \times 1.2}{6}$$

$$\frac{1}{4} \times \frac{20}{20} = \frac{20}{30} \text{(simple gearing)}$$

Now depends on gear ratio and pitch of the lead screw.

Table 8: Values of pitch of the lead screw and gear ratio.

Pitch of the lead screw (mm)	Gear ratio
6 mm	(0.25) 20:80/1:4
8 mm	(0.1875) 3:16
10 mm	(0.15) 3:20
12 mm	(.125) 1:8
14 mm	(0.107) 3:28

The expected outcome for the graduate is to impart knowledge and computation skill for gear ratio as the pitch of the lead screw changes.

Metal cutting process: thread cutting

Learning problem 8

Driver and the driven gear ratio change the efficiency and performance of the RPM of the mating parts. Determine and calculate the change in gears in an engine lathe to cut 12.5 threads per inch (T.P.I.), and the lead screw has 4.52 mm pitch for medium carbon steel.

Bloom's taxonomy cognitive level ACTION VERB: calculate

Expected outcome: knowledge (psychomotor domain) pertaining to skill development.

Learning solution 8

$$\text{Change in gear ratio} = \frac{\text{Driver gear(S)}}{\text{Driven gear(S)}} = \frac{\text{Pitch of the job in mm}}{\text{Pitch of the lead screw in mm}}$$

$$= \frac{1/12 \times 25.4 \text{ mm}}{4} = \frac{254}{12.5 \times 4.5 \times 10} = \frac{127 \times 20}{60 \times 4 \times 20}$$

$$\therefore \left[\text{Pitch} = \frac{1}{\text{No. of } \frac{\text{threads}}{\text{inch}}} \right]$$

$$\text{Gear ratio} = \frac{127}{80} \times \frac{20}{60} \text{ (compound gearing)}$$

Note: Assume single start thread; do not mention the no. of starts.
Depends on gear ratio and change in T.P.I.

Table 9: Values of speed and gear ratio.

Speed (threads per inch)	Gear ratio
12	0.529
14	0.453
16	0.396
18	0.352
20	0.3175

The expected outcome for the graduate is to impart knowledge and computation skill for gear ratio as T.P.I. changes.

Learning problem 9

Engine lathe machine tool is applicable for thread manufacturing for special fasteners. Calculate and determine the change gears for cutting 12.5 T.P.I. on an engine lathe having a lead screw of 4.5 T.P.I.

Bloom's taxonomy cognitive level ACTION VERB: determine and calculate

Expected outcome: knowledge (psychomotor domain) pertaining to skill development.

Learning solution 9

$$\textbf{Formula: } \text{Change in gear ratio} = \frac{\text{Driver gear(S)}}{\text{Driven gear(S)}}$$

$$= \frac{\text{Pitch of the job}}{\text{Pitch of the l.S.}}$$

$$= \frac{1/12.5}{1/4.5} = 4.22/12 = 1/3$$

$$1/3 \times 20/20 = \frac{20}{60} = 0.353$$

Depends on gear ratio and T.P.I.

Table 10: Values of thread per inch (T.P.I.) and gear ratio.

T.P.I. (inch)	Gear ratio
4	0.333
5	0.416
6	0.500
7	0.583
8	0.666

Thread per Inch versus Gear Ratio

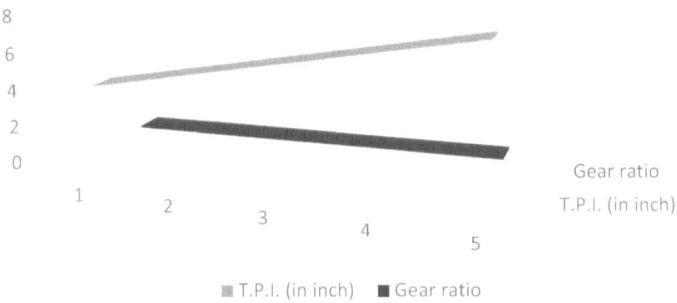

Figure 5: Graph between T.P.I. and gear ratio.

The expected outcome for the graduate is to impart knowledge and computation skill for gear ratio as T.P.I. changes.

Learning problem 10

Lathe machine tool is used to machine cylindrical workpiece materials using single point cutting tool or hard carbide inserts. Workpiece rotates fixed in a chuck, and cutting tool is fed in a direction perpendicular to the axis of the work material specimen. Determine and calculate the RPM, namely, workpiece (stainless steel) of 100.5–180.5 mm in diameter which is to be turned in a lathe machine tool to attend a cutting speed of 25.5 m/min.

Bloom's taxonomy cognitive level ACTION VERB: determine and calculate

Expected outcome: knowledge (psychomotor domain) pertaining to skill development.

Learning solution 10

$$\text{Cutting speed} = V = \frac{\pi D N}{1,000}$$

$$\text{Cutting speed, } V = 25 \, \text{m/min}$$

$$\text{Work specimen,} \quad D = 100 - 180 \, \text{mm}$$

$$25 = \frac{\pi \times 100.5 \, N}{1,000}$$

$$N = \frac{25.5 \times 1,000}{\pi \times 100}$$

$$N = 79.6 \, \text{RPM}$$

$$N = 80.2 \, \text{RPM}$$

Table 11: Values of diameter of workpiece and RPM.

Diameter	RPM
100	80
120	$66.31 \approx 70$
140	$56.84 \approx 60$
160	$49.73 \approx 50$
180	$44.20 \approx 40$

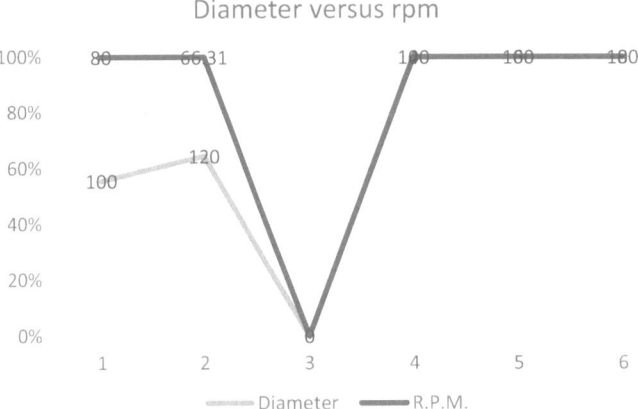

Figure 6: Graph between diameter and RPM.

The expected outcome for the graduate is to impart knowledge and computa-tion skill for change in diameter and RPM of chuck.

Learning problem 11

Facing is the first lathe machine operation before turning to remove the undesired sur-face on both sides of the faces of the cylindrical work material. Single point cutting tool (HSS) is used for the turning and facing of ductile materials. Calculate and determine the machining time for a facing operation on a lathe m/c for different facing lengths. Shift with 400.5 RPM and feed may be taken as 0.25 mm/rev for the steel-graded material.

Bloom's taxonomy cognitive level ACTION VERB: calculate

Expected outcome: knowledge (psychomotor domain) pertaining to skill development.

Learning solution 11

$$\text{Facing length,} \quad l = \frac{D}{2} = \frac{100}{2} = 50 \, \text{mm}$$

Job speed, $\quad N = 400.5$ RPM
Feed/rev, $\quad f = 0.2$ mm/rev
No. of cut $\quad = 1$
Diameter of job, $D = 100–180$ mm
Machining time

$$T_m = \frac{(A + l + O)}{fr \times N} \times \text{no. of cut}$$

$$T_m = \frac{(5 + 50 + 3)}{0.25 \times 400} \times 1$$

$$T_m = 0.725 \text{ min}$$

Now it depends on facing length (l) and machining time.

Table 12: Values of facing length and machine time.

Facing length (l)	Machine time (T_m)
50 mm	0.725
60 mm	0.85
70 mm	0.975
80 mm	1.1
90 mm	1.225

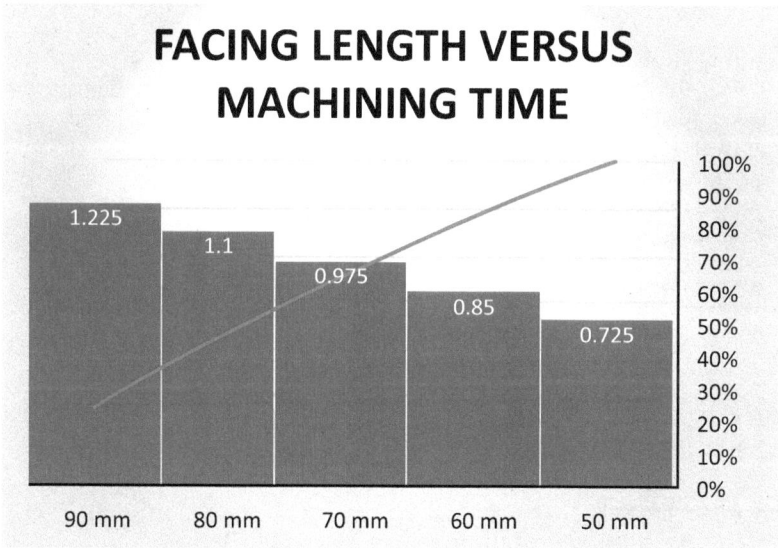

Figure 7: Graph between facing length and machining time.

The expected outcome for the graduate is to impart knowledge and computation skill for machining time as facing length changes.

Learning problem 12

Shaper is a conventional machine tool used in small-scale industries for flattening the plates, groove, and slot making including keyways. Calculate and determine the time required for a face of a rectangular block made of cast iron by a shaping metal cutting process.

  Data given:
  Length of rect. block = 400.5 mm
   Width of rect. block = 300.5 mm

Bloom's taxonomy cognitive level ACTION VERB: determine and calculate

Expected outcome: knowledge (psychomotor domain) pertaining to skill development.

Learning solution 12

$$\text{Length of stroke}, L = (l_a + l + l_o)/1,000 = \frac{20 + 400.5 + 10}{1,000} = 0.43\,\text{m}$$

$$\text{Time for cutting stroke}, t_c = \frac{L}{V_c} = \frac{0.43\,\text{m}}{10\;\text{m/min}} = 0.043\,\text{min}.$$

$$\text{Time for the return stroke}, t_r = \frac{L}{V_R} = \frac{0.43\,\text{m}}{15\;\text{m/min}} = 0.0287\,\text{min}.$$

Table 13: Values of shaping width and machining time.

b_o (mm)	T_m (min)
5	44.56
10	45.17
15	45.88
20	46.60
25	47.32

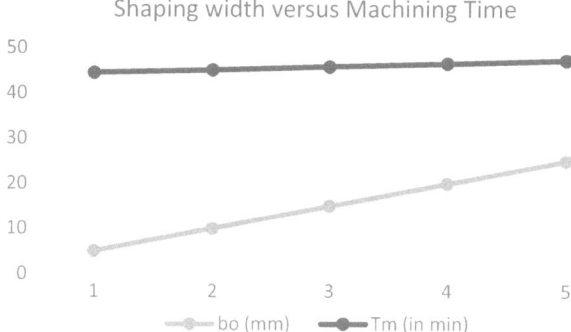

Figure 8: Graph between shaping width and machining time.

The expected outcome for the graduate is to impart knowledge and computation skill for shaping versus machining time observations.

Learning problem 13

Surface finish improvement on rough surfaces of the engineering materials like cast iron and mild steel can be gradually improved using shaping before grinding the faces of the plate. Calculate and determine the time required on a shaping m/c for complete one cut only on a plate mad of cast iron, namely, 200.5 mm × 300.5 mm; cutting speed is 15.5 m/min. The return time to cutting time ratio is 2:3, and the feed rate is 2.54 mm.

Bloom's taxonomy cognitive level ACTION VERB: determine and calculate

Expected outcome: knowledge (psychomotor domain) pertaining to skill development.

Learning solution 13

$$\text{Time taken to complete in double strokes} = \frac{L(1+m)}{1,000 \times V}$$

$$\text{Total no. of double strokes required to complete the job} = \frac{B}{5}$$

$$\text{Total time required to complete the cut} = \frac{C \times 3\ (1+m)}{1,000 \times V \times 5}$$

Table 14: Values of cutting ratio
and machining time.

Cutting ratio	T_m (min)
0.66	5.573
0.71	5.718
0.76	5.885
0.81	6.052
0.86	6.219

Cutting Time versus Machining Time

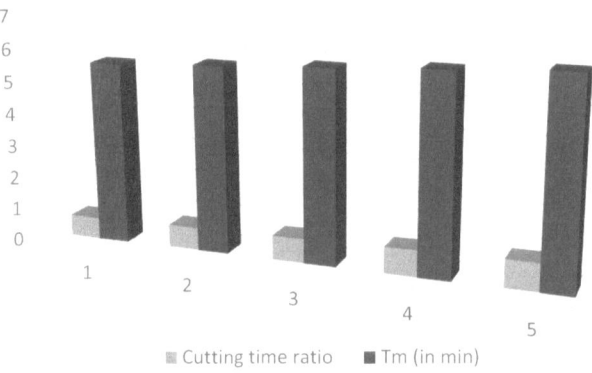

Cutting time ratio   Tm (in min)

Figure 9: Graph between cutting time and machining time.

The expected outcome for the graduate is to impart knowledge and computation skill for cutting ration versus machining time.

Learning problem 14

Grinders are used to improve the surface quality, namely, workpiece is pressed against the rotary grinder made of suitable reinforcement materials. Calculate and determine the wheel speed of a shaft/rod (aluminum 6065) with a diameter of 100.5 mm which is to be grinded.

Bloom's taxonomy cognitive level ACTION VERB: determine and calculate

Expected outcome: knowledge (psychomotor domain) pertaining to skill development.

Learning solution 14

Peripheral speed of the w/p sample $V_w = 15$ m/min
   Diameter of the w/p sample $D_w = 100.5$ mm
   Wheel speed, $N_w = ?$ RPM
   Formulae:

$$V_w = \frac{\pi D_w N_w}{1,000}$$

Let us find dependency of N_w on V_w.

Table 15: Values of peripheral speed and wheel speed.

V_w (m/min)	N_w (rpm)
15	48
20	64
25	80
30	96
35	112

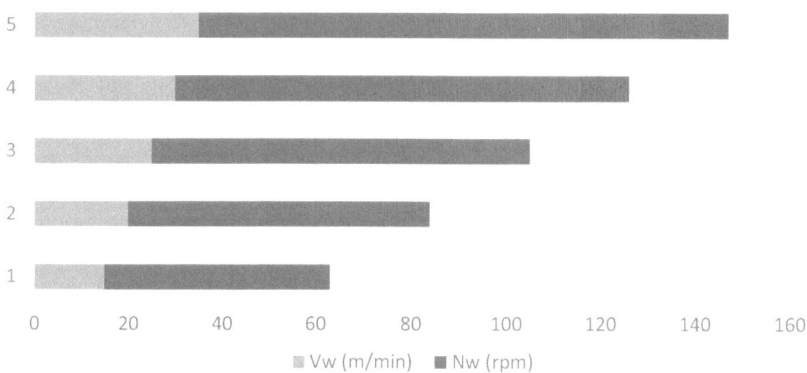

Figure 10: Graph between peripheral speed and wheel speed.

Metal cutting process: turning

The expected outcome for the graduate is to impart knowledge and computation skill for the grinding operation.

Learning problem 15

Accomplish turning operation and apply cylindrical parts using single point cutting tool on the lathe machine tool. Feed and depth of cut are set as per the need of the diameter to be turned and surface quality. Evaluate and determine the machining time for turning of aluminum (6063) with 50.5 mm diameter brass bar to 42.5 mm diameter over length of 50 mm and pin speed of 450RPM. The maximum depth of cut is limited to 3.2 mm and feed f is set at 0.22 mm/rev.

Bloom's taxonomy cognitive level ACTION VERB: determine and calculate

Expected outcome: knowledge (psychomotor domain) pertaining to skill development.

Learning solution 15

$$T_m = \frac{(L + l_o)}{fN}$$

where we took N different values

$$N = 350\text{--}450 \text{ RPM}$$

$$T_m = \frac{(L + l_a)}{fN} = \frac{2(50.5 + 3)}{0.2 \times 350} = 1.514 \text{ min}$$

$$T_m = \frac{2 \times (50 + 3)}{0.2 \times 410} = 1.292 \text{ min}$$

$$N = 370 \text{ RPM}$$

$$T_m = \frac{2 \times (50 + 3)}{0.2 \times 370} = 1.432$$

Table 16: Values of pin speed and machining time.

N	T_m
350	1.514
370	1.432
390	1.358
410	1.292
430	1.232
450	1.177

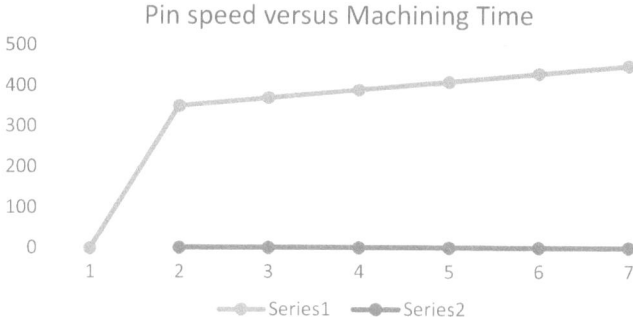

Figure 11: Graph between pin speed and machining time.

The expected outcome for the graduate is to impart knowledge and computation skill for speed variation and machining time required.

Learning problem 16

Twist drill cutting tools (having more than one edges with flutes) are used for manufacturing hole on engineering materials. About 12.5 mm diameter hole is to be drilled in steel to a depth of cut of 50.5 mm with HSS drills. Determine and calculate the time of drilling 100 pieces, if the (aluminum 6062) setup time is 30 s.

Bloom's taxonomy cognitive level ACTION VERB: determine and calculate

Expected outcome: knowledge (psychomotor domain) pertaining to skill development

Learning solution 16

$$\text{Drill RPM} = \frac{30 \times 1,000}{(\pi \times 12.5)} = 79.33$$

The available speed on the machine nearest to the above value is 750 RPM.
    Time required for drilling hole up to 50 mm depth

$$= \frac{50 \times 60}{0.15 \times 750} = 26.66 \text{ s} \approx 27 \text{ s}.$$

Time calculated for removal of chips in between = 5 s.

Table 17: Values of RPM and drilling time.

N (RPM)	Time for drilling T (s)
650	30.76
670	29.84
690	28.97
710	28.16
730	27.39
750	27.66

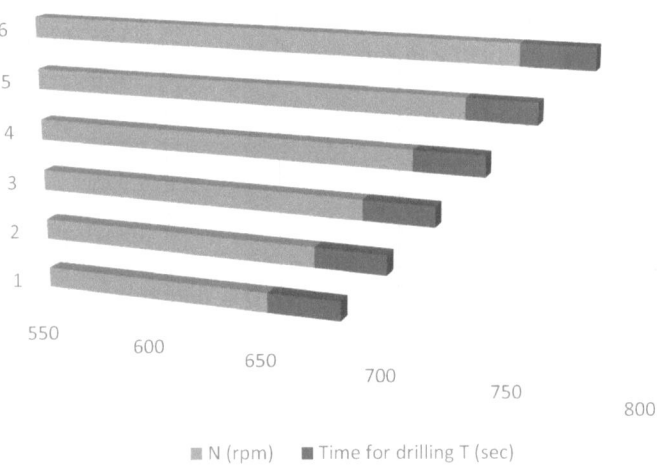

Figure 12: Graph between RPM and drilling time.

The expected outcome for the graduate is to impart knowledge and computation skill for drilling time required with respect to change in RPM.

Learning problem 17

Gear manufacturing is important for transmit power/motion to machine elements. Determine and calculate the machining time of milling process (gear manufacturing) for Al-6063 engineering material its to be cut. Tool diameter is 100.5 mm with 12 cutting teeth, and the depth of cut is 5.5 mm.

Bloom's taxonomy cognitive level ACTION VERB: determine and calculate

Expected outcome: knowledge (psychomotor domain) pertaining to skill development

Learning solution 17

Time length formula

$$= 2\sqrt{(5 \times (100.5 - 5.5))}$$

$$= 43.58$$

Thus, $\dfrac{N = 30 \times 100}{(\pi \times 100)} = 92.45\,\text{RPM}$

Let us take that 90 RPM is available on the machine feed per min = $90 \times 12 \times 0.22 = 216$ mm/min

Time for cutting on both $= \dfrac{90.5}{216}\,\text{min}$

Time for indexing rapid reversing = 0.25 min

Time for machining the gear = $(90/216 + 0.254) = 40$ min.

Table 18: Values of feed rate and machining time.

Feed (mm per min)	Time (min)
50	125
100	69
150	51
200	42
250	36.6

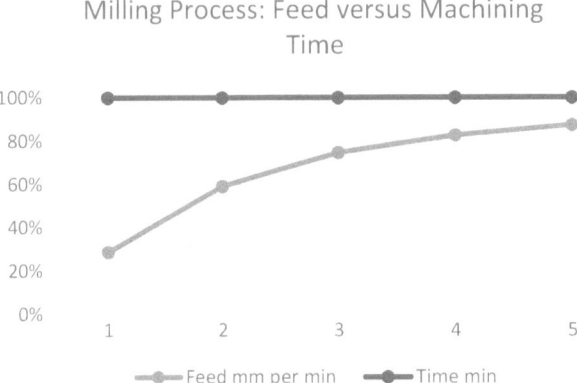

Figure 13: Graph between feed rate and machining time.

The expected outcome for the graduate is to impart knowledge and computation skill feed versus machining time required.

Learning problem 18

Orthogonal and oblique cutting are the variant methods applied for cylindrical shaped engineering materials. Al-6061 is turned on end in orthogonal cutting condition with a tool of rake having 20°. Chip length of 85.5 mm is obtained during metal cutting practice from an uncut chip length of 202 mm and cutting with a depth of cut of 0.5 mm. Determine and calculate the thickness (chip) and variation of shear angle with respect to the rake angle.

Bloom's taxonomy cognitive level ACTION VERB: determine and calculate

Expected outcome: knowledge (psychomotor domain) pertaining to skill development.

Learning solution 18

Chip thickness ratio $= 85.5/202 = 0.42$

$$r_t = \frac{t}{t_c} = \frac{V_c}{V} = \frac{L_c}{L}$$

Shear angle $\varnothing$ is to determine

$$\emptyset = \tan^{-1}\left(\frac{r_t \cos\alpha.}{1 - r_t \sin\alpha}\right) = \tan^{-1}\left(\frac{0.42 \times \cos 20}{1 - 0.42 \times \sin 20}\right)$$

$$= 24.74°$$

Chip thickness $t_c = t/0.42 = 1.18\,\text{mm}$

$$t_c = \frac{5}{0.42} = 1.28\,\text{mm}$$

Table 19: Values of rake angle and shear angle and wheel speed.

α (rake angle)	Ø (shear angle)
0	22.78
5	23.47
10	24.04
15	24.47
20	24.74

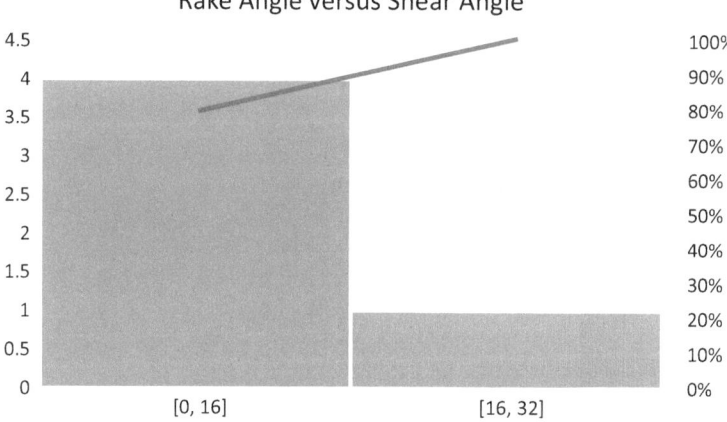

Figure 14: Graph between rake angle and shear angle.

The expected outcome for the graduate is to impart knowledge and computation skill for turning operation to verify the effect of change of shear angle and rake angle.

Learning problem 19

Tool geometry has a relationship with rake and shear angles. Calculate and determine the relationship variation of chip thickness and shear angle machining of a component on shaper machining tool. The work material (cast iron) having length along the stork is 100.5 mm and the chip length of 40.5 mm is obtained with a tool of 15° rake angle.

Bloom's taxonomy cognitive level ACTION VERB: determine and calculate

Expected outcome: knowledge (psychomotor domain) pertaining to skill development.

Learning solution 19

$$t \times l = t_c \times l_c$$

Therefore, chip thickness ratio $t/t_c = l_c/L = 40.5/100.5 = 0.4$

$$\emptyset = \tan^{-1}\left(\frac{r \cos \alpha}{l - r \sin \alpha}\right)$$

$$= \tan^{-1}\left(\frac{0.4 \times \cos 15°}{1.04 \times \sin 15°}\right) = 23.335$$

Chip thickness $= t_c = t/r$

$$= 1.5/0.4 = 3.73 \, \text{mm}$$

if r values change.

Table 20: Values of chip thickness and shear angle.

r (chip thickness)	$\emptyset$ (shear angle)
0.1	5.66
0.2	11.51
0.3	17.44
0.4	23.31
0.5	29.01

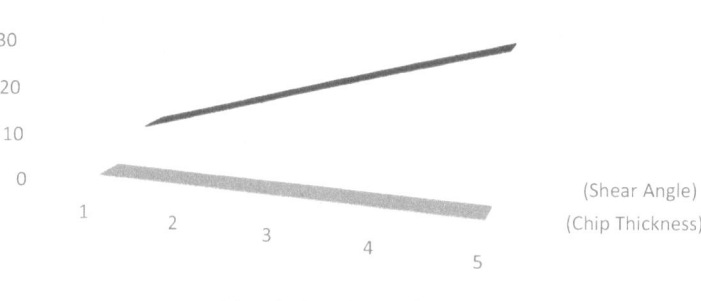
Chip Thickness verus Shear Angle

Figure 15: Graph between chip thickness and shear angle.

The expected outcome for the graduate is to impart knowledge and computation skill for turning and for predicting the relationship, namely, chip thickness and SPCT shear angle.

Learning problem 20

Two-dimensional cutting or orthogonal cutting operation for metal cutting of ductile materials is accomplished with a work and cutting tool to be aligned perpendicular to the feed movement. Using the input data given below, calculate and determine the variation of shear angle and force:

Rake angle of tool = 15.5°
Uncut chip thickness t = 0.255 mm
Width of chip = 2.1 mm
Chip thickness ratio r = 0.46
Fraction angle β = 40°

Bloom's taxonomy cognitive level ACTION VERB: determine and calculate

Expected outcome: knowledge (psychomotor domain) pertaining to skill development.

Learning solution 20

$$\tan \varnothing = \frac{r \cos \propto}{1 - r \sin \propto} = \frac{0.46 \cos 15.5°}{(1 - 0.46 \sin 15)} = 26.76$$

$$F_s = \text{t.b.k}/\sin\varnothing = \frac{0.255 \times 2 \times 250}{\sin 26.76} = 279.62$$

$$\text{force} = \frac{277.62}{\cos(\varnothing + \beta - \propto)} = \frac{277.62}{\cos 51.76} = 458.52\,\text{N}$$

$$\text{Cutting force component} = \frac{448.52}{\cos(25°)} = 406.85\,\text{N}$$

Table 21: Values of shear angle and force components.

$\varnothing$ (degree)	F_s (N)
5	1434.2
10	719.8
15	482.96
20	365.47
25	295.77

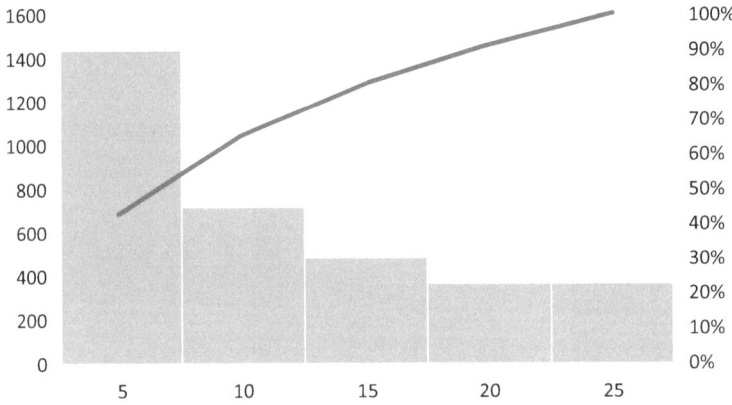

Figure 16: Graph between shear angle and force component.

The expected outcome for the graduate is to impart knowledge and computation skill for shear angle variation with force.

Chapter 2
Metal cutting analysis

Learning problem 21

Lathe machining tool (conventional/CNC based) is used for various machining operations such as facing, turning, taper turning and threading. Calculate the force and shear angle for the machining of mild steel work specimen, considering length of cut chip obtained from uncut chip length of 100 mm = 50.5 mm, rake angle of tool = 10°, uncut thickness = 0.25 mm, and width of cut = 1.5 mm.

Bloom's taxonomy cognitive level ACTION VERB: determine and calculate

Expected outcome: knowledge (psychomotor domain) pertaining to skill development.

Learning solution 21

$$\text{Chip thickness ratio} = \frac{L_c}{L} = 0.5$$

$$\text{Shear angle} = \frac{r\cos\varnothing}{1 - r\sin\varnothing} = \frac{0.5\cos 10}{1 - 0.5\sin 10} = 26.34°$$

$$(F_5)\text{ Force along the shear plane} = \frac{t.b.k}{\sin\varnothing} = 123.4\text{ N}$$

$$\text{Resultant force on the cutting tool }(R) = \frac{F_5}{\cos(\varnothing + \beta + \propto)} = 212.08\text{ N}$$

$$R\text{ cutting force component}(F_h) = R \times \cos(\beta - 2) = 203.64\text{ N}$$

$$u = \tan\beta = \infty \ \ \beta = \tan^{-1}(0.8) = 39.65$$

Table 22: Values of friction angle and force.

Friction angle (degree)	Force R (N)
0	133.16
10	143.61
20	161.15
30	190.15
40	240.8

https://doi.org/10.1515/9783110676662-002

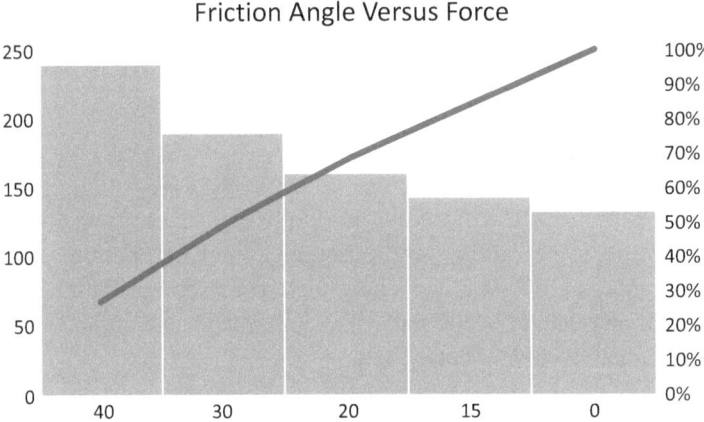

Figure 17: Graph between friction angle and force.

The expected outcome for the graduate is to impart knowledge and computation skill for friction angle versus force.

Learning problem 22

Feed, depth of cut, and RPM of the moving cylindrical part (work specimen) play an important role for detecting the machining performance outcome variables. Al-6063 bar 80.5 min diameter with cutting speed of 60 m/min is turned. The feed and depth of cut are set at 0.45 mm/rev and 3.51 mm. Determine and calculate the specific resistance and the unit power of material.

Bloom's taxonomy cognitive level ACTION VERB: determine and calculate

Expected outcome: knowledge (psychomotor domain) pertaining to skill development.

Learning solution 22

Cross section of Al-6063 being removed $= f \times d$

$$= 0.45 \times 3.51 = 1.42 \text{ mm}^2$$

$$\text{Specific cutting resistance} = f_c/f_d = \frac{750}{1.4} \text{ N/mm}^2$$

$$\text{Power being consumed} = \frac{750 \times 60}{(1,000 \times 6)} = 0.72 \text{ kW}$$

$$\text{Material being removed per second} = \frac{1.4 \times 60 \times 1,000}{(60 \times 1,000)}$$

$$= \frac{fdV}{60}\,\text{cm}^3/\text{s}$$

$$\text{Unit power} = \frac{0.75}{1.4} = 0.537\,\text{kW}/\text{cm}^3/\text{s}$$

$$\text{where Up} = \frac{F_c}{1,000 \times f \times d}$$

Table 23: Values of feed rate and unit power.

Feed	Unit power
0.1	2.14
0.2	1.07
0.3	0.713
0.4	0.535
0.5	0.428

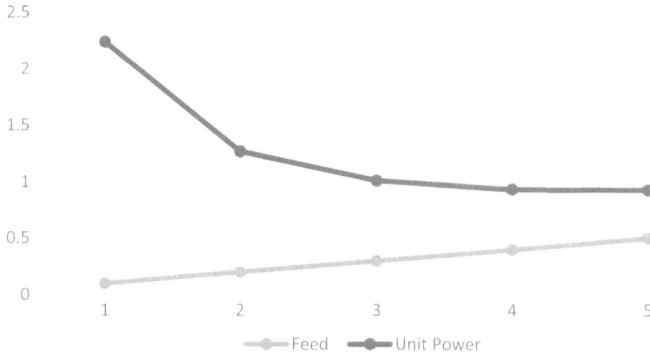

Feed Versus unit power

Figure 18: Graph between feed rate and unit power.

The expected outcome for the graduate is to impart knowledge and computation skill for feed rate and unit power obtained.

Learning problem 23

Drilling and its variant metal cutting operations like boring are used to make a hole of specified geometry and dimensions for automobile and fabrication machine tool parts. Calculate and determine the trust/torque required for drilling 20.5 mm diameter holes in Al-6061 rectangular plate of 25.5 mm thickness. The feed is set at 0.255 mm/rev to take cutting speed of 40.5 m/min.

Bloom's taxonomy cognitive level ACTION VERB: determine and calculate

Expected outcome: knowledge (psychomotor domain) pertaining to skill development.

Learning solution 23

Trust force

$$F = 196.2 f^{1.1} D^{1.2} N$$

$$= 196.2 \, (0.15)^{1.1} (20.5)^{1.2}$$

$$= 196.2 \times 0.124 \times 36.4$$

$$= 883.84 \, N$$

Torque

$$M = 0.1265 f^{0.83} D^{1.9} \, Nm$$

$$= 0.1265 \times (0.15)^{0.83} (20.5)^{1.9}$$

$$= 0.1265 \times 0.207 \times 295.45$$

$$= 7.862 \, Nm$$

Table 24: Values of feed and force.

f (feed)	F_N (force)
0.03	150.88
0.06	323.41
0.09	505.04
0.12	693.26
0.15	885.84

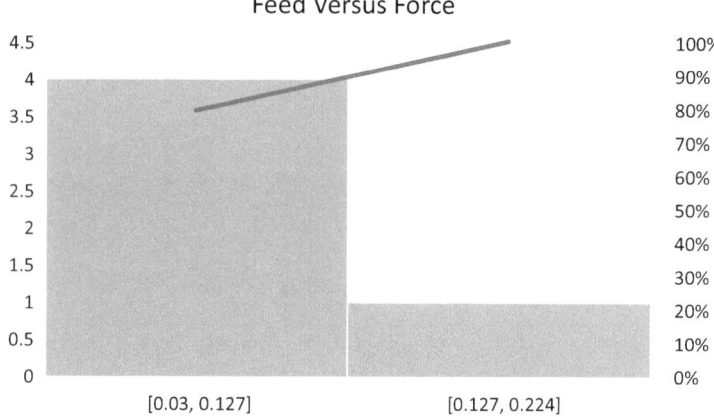

Figure 19: Graph between feed rate and force.

The expected outcome for the graduate is to impart knowledge and computation skill for feed rate versus force calculation.

Learning problem 24

Drilling involves two principal cutting edges made of HSS and other carbide-coated tools. Calculate and determine the trust force, torque, and power required to drill Al-6063 with drill cutting tool of 14.5 mm diameter. The feed = 0.24 mm/rev and rotational speed is at 410 RPM.

Bloom's taxonomy cognitive level ACTION VERB: determine and calculate

Expected outcome: knowledge (psychomotor domain) pertaining to skill development.

Learning solution 24

$$F = 67.4 \times 10^3 \, f^{0.87} (D/127 + W/D)^{1.9} \text{ N}$$

$$= 67.4 \times 10^3 \, (0.2)^{0.87} (14.5/127 + 0.153)^{1.9} \text{ N}$$

$$= 1169.8 \text{ N}$$

Torque

$$M = 0.292 \, f^{0.6} D^{1.7} \text{ Nm}$$

$$= 0.292 (0.24)^{0.6} (12)^{1.7}$$

$$= 7.9 \text{ Nm}$$

$$\text{Rotational speed} = 300 \times \frac{2\pi}{60} \text{ rev/s}$$

$$\text{Power consumed in drilling} = \frac{7.6 \times 410 \times 2\pi}{60} = 245.76 \text{ W}$$

Table 25: Values of RPM and power.

N(pm)	Power
60	47.75
120	95.50
180	143.25
240	191.00
300	238.76

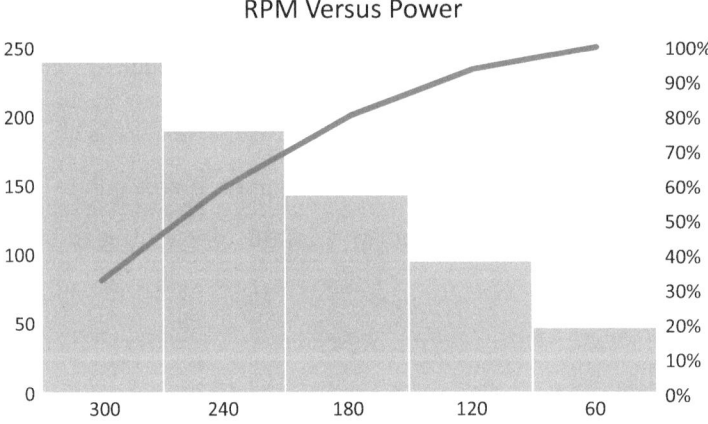

Figure 20: Graph between RPM and power.

The expected outcome for the graduate is to impart knowledge and computation skill for RPM versus power.

Learning problem 25

Bake rake angle guides the direction of chip flow and protect the cutting edge. Angle may be positive, negative, or neutral. Determine and calculate the rake angle for the cylindrical work specimen (mild steel) with back rack angle side equal to 10° and side cutting edge angle 30°. The feed and depth of cut are set at 0.24 mm/rev and 3.5 mm, respectively.

Bloom's taxonomy cognitive level ACTION VERB: determine and calculate

Expected outcome: knowledge (psychomotor domain) pertaining to skill development.

Learning solution 25

$$\gamma_s = \text{side cutting edge angle}$$

$$\tan \varphi = \frac{d}{(f + d \tan \gamma_1)}$$

$$\text{Chip flow angle } (\varphi) = \tan^{-1}(3/(0.24 + 3.5 \times \tan 30))$$

$$= 57.21°$$

$$\alpha_c = \tan^{-1}(\tan \alpha_b \cos \varphi + \tan \alpha_s \sin \varphi)$$

$$\alpha_c = \tan^{-1}(\tan 10° \cos 57.21° + \tan 10° \sin 57.21°)$$

$$= \tan^{-1}(0.176 \, (0.415 + 0.840))$$

$$= 13.8°$$

Table 26: Values of side cutting edge angle and shear angle.

γ_s (side cutting edge angle)	φ (shear angle)
6	80.25
12	74.39
18	68.61
24	62.89
30	57.21

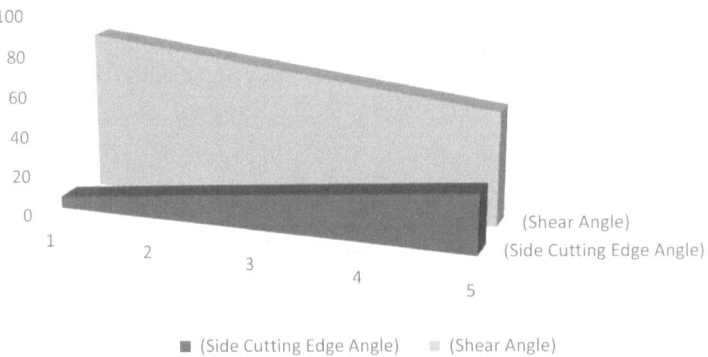

Figure 21: Graph between SCEA and shear angle.

The expected outcome for the graduate is to impart knowledge and computation skill for shear angle versus side cutting edge angle.

Learning problem 26

High temperature is established at the chip tool interface area in the form of spot welds. Calculate and determine the temperature rise at the shear plane of a steel-graded alloy of yield strength in shear of 310 N/mm² with a tool of 12° rack angle. The fraction angle $\beta = 44$.

Bloom's taxonomy cognitive level ACTION VERB: determine and calculate

Expected outcome: knowledge (psychomotor domain) pertaining to skill development.

Learning solution 26

$$\varnothing = 45 - 1/2(44 - 2) = 23°$$

$$E_s = \frac{K \cos \propto}{\sin \varnothing . \cos(\varnothing - \propto)}$$

$$E_s = \frac{310 \times \cos 10}{\sin 29 \times \cos 19} = 644.5 \, \text{Nm/cm}^3$$

$$\text{Temperature rise at a shear plane} = \frac{664.5}{(0.44 \times 7.87)}$$

$$= 186.12\,°C$$

Table 27: Values of rake angle and temperature.

$\propto$ (rake angle)	T (temperature rise)
2	222.53
4	215.80
6	203.28
8	194.46
10	186.12 °C

The expected outcome for the graduate is to impart knowledge and computation skill for rake angle versus temperature rise.

Learning problem 27

Al-6061 rod of 50.5 mm diameter is to be turned over a length of 160 mm with a depth of cut of 1.55 mm feed of 0.2 mm/rev at 230 RPM by the HSS tool. Determine and calculate the required parts to be used for turning operation.

Bloom's taxonomy cognitive level ACTION VERB: determine and calculate

Expected outcome: knowledge (psychomotor domain) pertaining to skill development.

Learning solution 27

$$\text{Cutting speed } (V) = \frac{\pi\,D\,N}{1,000} = \frac{\pi \times 50.5 \times 230}{1,000} = 36.13\,\text{m/min}$$

Table 28: Values of depth of cut and machining time.

D (mm), depth of cut	T (min), time
00.3	116.88
00.6	77.11
00.9	60.46
01.2	50.87
01.5	44.5

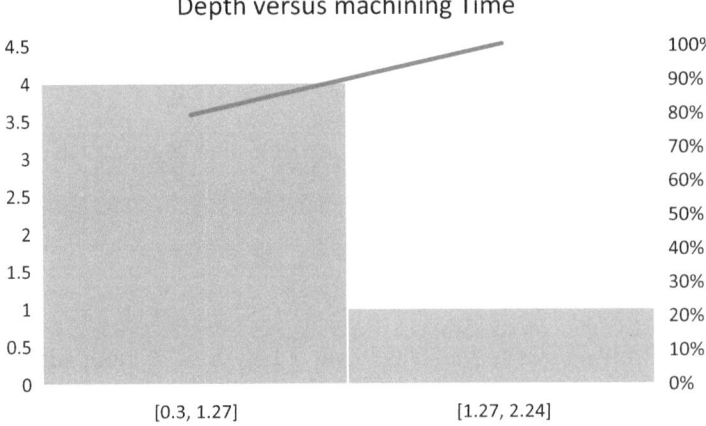

Figure 22: Graph between depth and machining time.

The expected outcome for the graduate is to impart knowledge and computation skill for depth of cut versus machining time.

Learning problem 28

Merchant force circle diagram and analysis on various force components are predicted with chip thickness ratio and the contact area and length. Calculate and determine the contact length.

Uncut chip thickness = $t = 0.25$ mm

Cutting force component in the direction cutting velocity (F_h) = 125 N

Cutting force component normal to the machined surface = F_v = 65 N

Chip thickness ratio = $r_t = 0.45$ rake angle = $\propto$ = 12

Width of cut $b = 3.5$ m

Bloom's taxonomy cognitive level ACTION VERB: determine

Expected outcome: knowledge (psychomotor domain) pertaining to skill development.

Learning solution 28

The shear plane angle is given by

$$\varnothing = \tan^{-1}\left[\frac{0.45\cos 12}{1 - 0.45\sin 12}\right] = \tan^{-1}(0.48)$$

$$\varnothing = 25.64$$

Now

$$\tan(\beta - \alpha) = \frac{f_v}{f_n} = \frac{60}{120} = 0.5$$

$$\beta - \alpha = 26.56$$

$$\beta = 36.56°$$

The natural contact length l_n is obtained using

$$h_n = \left[\frac{t.\sin[\pi/4 + 1/2(\beta - 2)]}{\sin[\pi/4 - 1/2(\beta - 2)\cos\beta]}\right]$$

$$l_n = \left[\frac{0.2\sin(25.64 + 36.56 - 10)}{(\sin 25.64 \cos 36.56)}\right]$$

$$= 0.454 \text{ mm}$$

Table 29: Values of friction angle and contact length.

β (degree)	l_n (mm)
10	0.203
15	0.225
21	0.259
28	0.311
36.56	0.454

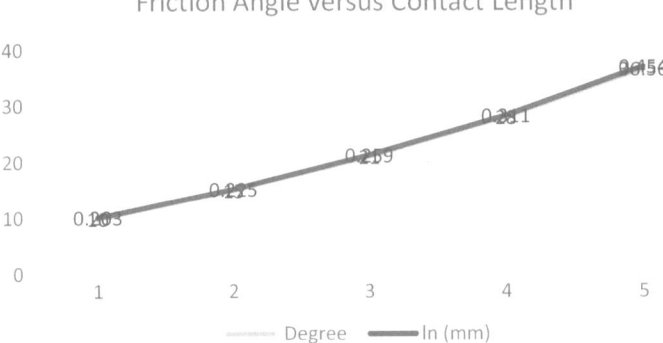

Figure 23: Graph between friction angle and contact length.

The expected outcome for the graduate is to impart knowledge and computation skill for friction angle versus contact length.

Learning problem 29

Surface grinding metal cutting process is a secondary process needed for improving texture and surface reliability/quality. Calculate and determine the uncut chip thickness for a cast iron plate by an SiC wheel of diameter 150 min under the following conditions:
a) No. of active grits per unit length along the wheel surface = 22 mm
b) Depth/feed = 40 mm

Bloom's taxonomy cognitive level ACTION VERB: determine

Expected outcome: knowledge (psychomotor domain) pertaining to skill development.

Learning solution 29

Uncut thickness; $a_{\mathrm{ng}} = \dfrac{V_{\mathrm{N}}}{MV_{\mathrm{g}}} \left(\dfrac{d}{D_y}\right)^{1/2}$, $a_{\mathrm{ang}} = \dfrac{2}{22 \times 3,000} \left(\dfrac{0.04}{150}\right)^{1/2}$

Table 30: Values of wheel velocity and uncut thickness.

S. no.	V_w (m/min)	a_{ang} (mm)
1	1.6	0.00435
2	1.7	0.00462
3	1.8	0.00489
4	1.9	0.00517
5	2	0.0054

The machined time $(t_m) = \dfrac{2\,I_m}{f \times n_w} = \dfrac{2 \times 53}{0.2 \times 450} = 1.20\ \text{min}$

Table 31: Values of feed rate and machining time.

Feed	Machining time (t_m)
0.04	5.888
0.08	2.9444
0.12	1.96296
0.16	1.4722
0.20	1.1778

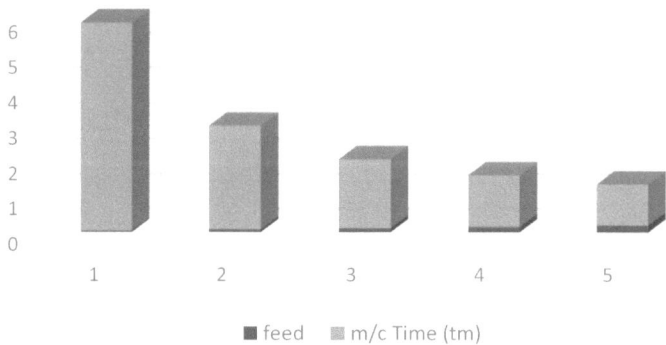

Feed versus Machining Time

■ feed ■ m/c Time (tm)

Figure 24: Graph between feed rate and machining time.

The expected outcome for the graduate is to impart knowledge and computation skill for feed versus machining time.

Learning problem 30

Aluminum-6063 rod is widely used in automobile and fabrication industries. HSS cutting tool is fed and the unwanted materials in the form of chips are generated during rotation of work material via a chuck bar. Determine the tool life for cutting velocity of 42 m/min, if the tool life is $VT^{0.2} = 84$.

Bloom's taxonomy cognitive level ACTION VERB: determine

Expected outcome: knowledge (psychomotor domain) pertaining to skill development.

Learning solution 30

$$T = (84/42)^{1/0.2} \cong 22 \text{ min}$$

The cutting speed for 60 min tool life is given by

$$V_{T=60} = 80/(60)^{0.2} = 35.27 \text{ m/min}$$

Table 32: Values of time and cutting speed.

T (min)	V (m/min)
10	50.47
20	43.94
30	40.51
40	38.25
50	36.58
60	35.27

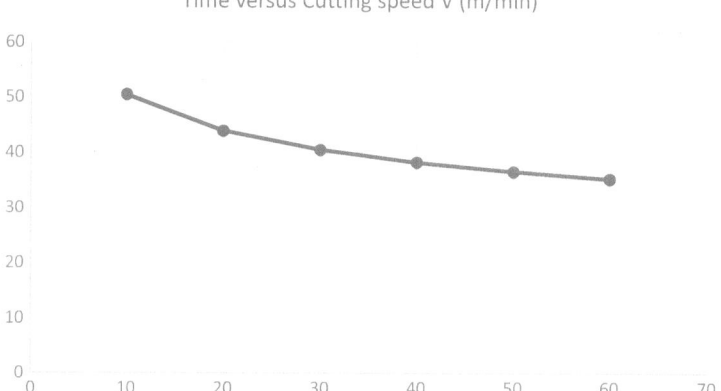

Figure 25: Graph between time and cutting speed.

The expected outcome for the graduate is to impart knowledge and computation skill cutting speed versus machining time.

Learning problem 31

Cutting temperature can be analyzed during experimental investigations and through simulation analysis for the machining of hard-to-machine materials. Calculate and determine the average cutting temperature to change by two times the cutting velocity and reducing the principal cutting edge angle from 60° to 30° for Ti–6Al–4V alloy used for aerospace applications.

Bloom's taxonomy cognitive level ACTION VERB: determine and calculate

Expected outcome: knowledge (psychomotor domain) pertaining to skill development.

Learning solution 31

Avg. cutting temperature $T_{avg} \propto \sqrt{V_c S_o \sin \theta}$

"θ" is available; $V_c \rightarrow$ cutting velocity

$S_o \rightarrow$

$$\frac{T_{avg2}}{T_{avg1}} = \sqrt{\frac{2V_c S_o \sin \theta}{V_c S_o \sin 90}} = \sqrt{2 \sin \theta}$$

$$\frac{T_{avg2}}{T_{avg1}} = \sqrt{2\sin 30°} = 1 \quad \rightarrow \quad \text{change} = 0\,°\text{C}$$

Table 33: Values of PCEA and change of temperature.

θ (PCEA)	Change (%)
30	0
42	15.67
54	27.2
66	35.17
78	39.86

PCEA Versus Change in percentage

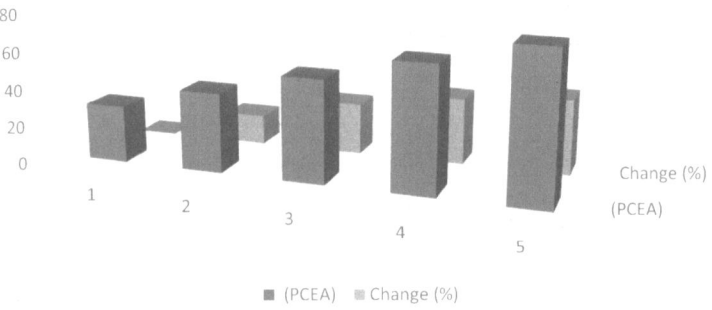

Figure 26: Graph between PCEA and change of temp. (%).

The expected outcome for the graduate is to impart knowledge and computation skill for PCEA and change in percentage of temperature.

Learning problem 32

Surface roughness of die-steel needs to improve after turning operation on lathe machine tool. Calculate and determine the surface roughness at feed of 0.4 mm/rev if the tool's cutting angles are 40° and 15°.

Bloom's taxonomy cognitive level ACTION VERB: determine and calculate

Expected outcome: knowledge (psychomotor domain) pertaining to skill development.

Learning solution 32

Surface roughness,

$$h_{\max} = \frac{S_0}{C_0 + \emptyset + C_0 + \emptyset\emptyset} = 0.0928\,\text{mm} = 92.8\,\text{Hm}$$

$$h_{\max} = \frac{0.4}{\cot 40° + \cot 15°}$$

"∅" is variable.

Table 34: Values of cutting angle and surface roughness.

∅ (degree)	$h_{\max}$ (µ m)
12	47.4
24	70.4
36	78.3
48	86.3
60	92.8

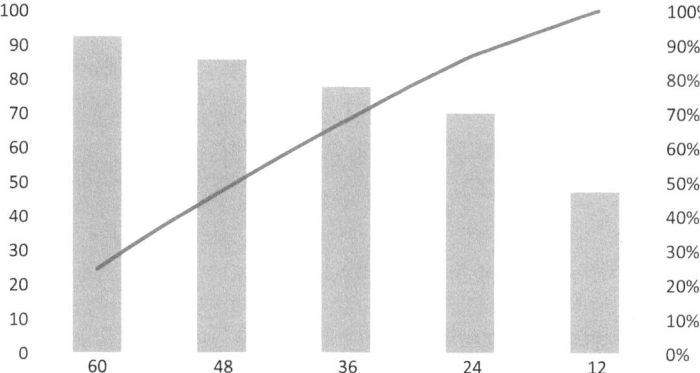

Figure 27: Graph between cutting angle and surface roughness.

The expected outcome for the graduate is to impart knowledge and computation skill for shear angle and surface roughness.

Cutting process single point cutting tool

Learning problem 33

Tool signature of the single point cutting tool is designated by seven different angles and nose radius. The design is recommended to achieve the maximum efficiency of metal cutting during turning on the lathe machine tool. Calculate and determine the orthogonal rake angle (r_o) of the cutting tool (HSS) specified in the ASA system as 12°, –10°, 8°, 6°, 15°, 30°, 0 (mm).

Bloom's taxonomy cognitive level ACTION VERB: determine and calculate

Expected outcome: knowledge (psychomotor domain) pertaining to skill development.

Learning solution 33

Orthogonal rake angle (r_o)

$$\tan r_o = \tan r_x \sin \varnothing + \tan r_y . \cos \varnothing$$

$$r_x = 10^0, r_y = 12^0, \varnothing = 90 - 30^0 = 60^0$$

"$\varnothing$" is variable

$$\tan r_o = \tan(12), \sin(60) + \tan(-10). \cos 60$$

$$r_o = 3.68°$$

Table 35: Values of shear angle and orthogonal rake angle.

$\varnothing$ (degree)	Orthogonal rake angle r_o (degree)
12	−1.132
24	−0.48
36	0.576
48	1.988
60	3.68

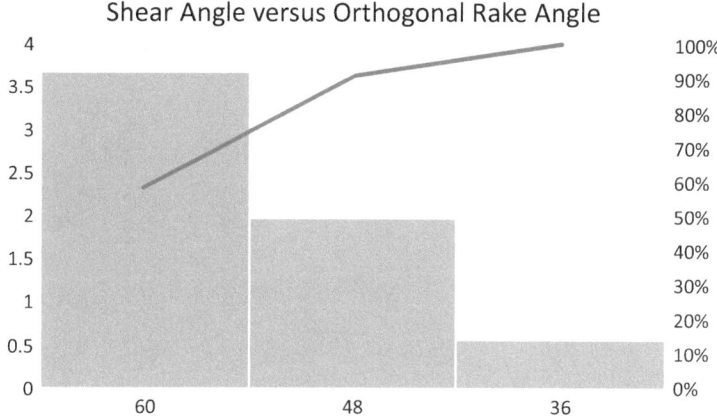

Figure 28: Graph between shear angle and orthogonal rake angle.

The expected outcome for the graduate is to impart knowledge and computation skill for shear angle versus orthogonal rake angle.

Learning problem 34

Clearance angles reduce the pressure and avoid the rubbing of work material with tool edge, preventing larger friction and heat generation during the metal cutting practices. Calculate and determine the values of side clearance angle of HSS tool, with tool geometry designated as −12°, 12°, 8°, 6°, 15°, 75°, 0 (mm) in orthogonal rake system.

Bloom's taxonomy cognitive level ACTION VERB: determine

Expected outcome: knowledge (psychomotor domain) pertaining to skill development.

Learning solution 34

Side clearance angle (α_x)

$$\cot \alpha_x = \cot \alpha_0 \sin \varnothing - \tan \lambda \cos \phi$$

$$\alpha_0 = 8.00 \quad \phi = 75° \quad \lambda = 12$$

$\varnothing$ is variable

$$\cot \alpha_x = \cot (8.00). \sin(75) - \tan (-12) \cos 75 = 7.918$$

$$\rightarrow \alpha_x = 8.22$$

Table 36: Values of side clearance angle and tool designate.

Ø (degree)	Side clearance angle α_x (degree)
15	26.43
30	15.08
45	10.976
60	9.09
75	8.22

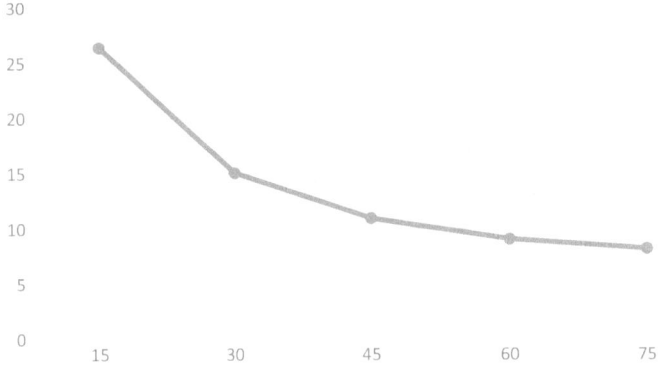

Figure 29: Graph between side clearance angle and tool designate.

The expected outcome for the graduate is to impart knowledge and computation skill for shear angle relationship with side clearance angle.

Learning problem 35

Twist drills are available in various types, namely, center drill, step drill, taper drill, and slot drill for hole manufacturing. Determine and calculate the axial rake angle for a twist drill whose helix angle (θ) is 34° and the chisel edge diameter is 4 mm.

Bloom's taxonomy cognitive level ACTION VERB: determine

Expected outcome: knowledge (psychomotor domain) pertaining to skill development to machine a taper cut on cylindrical shaft

Learning solution 35

$$\tan(r_{Di}) = \left(\frac{r_i}{6}\right)\tan\theta$$

$$= \left(\frac{5}{10}\right)\tan 34 = 0.4524$$

$$\rightarrow r_{Di} = 18.25°$$

"θ" is variable.

Table 37: Values of axial rake angle for twist drill and helix angle.

θ (degree)	r_{Di} (degree)
8	4.02
14	7.106
20	10.314
26	13.705
32	17.35

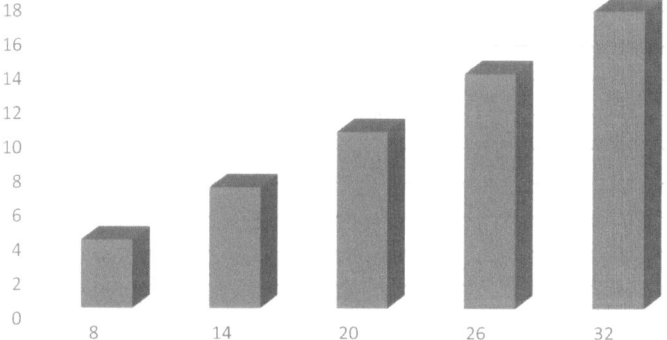

Figure 30: Graph between axial rake angle for twist drill and helix angle.

The expected outcome for the graduate is to impart knowledge and computation skill for axial rake angle of twist drill for hole manufacturing.

Learning problem 36

Calculate and determine the side rake angle made of HSS engineering material for machining, namely, turning of die steel maintaining a pitch value of 1.5 mm. Tool signature: 0°, 0°, 10°, 30°, 30°, 0 (mm).

Bloom's taxonomy cognitive level ACTION VERB: determine and calculate

Expected outcome: knowledge (psychomotor domain) pertaining to skill development.

Learning solution 36

$$r_{xw} = \tan^{-1}\left(\frac{S_0}{2\pi r}\right) + r_{xD},$$

$$S_0(Pitch) = 1.5mm \; r_{xD} = 0^0$$

$$r = 8mm$$

S_0 variable

$$r_{xw} = \tan^{-1}\left(\frac{1.5}{2\pi(18)}\right) = 3.28^\circ$$

Table 38: Values of pitch and surface finish.

S_0 (mm)	r_{xw} (degree)
0.4	0.456
0.8	0.9116
1.2	1.367
1.6	1.822
2	2.28

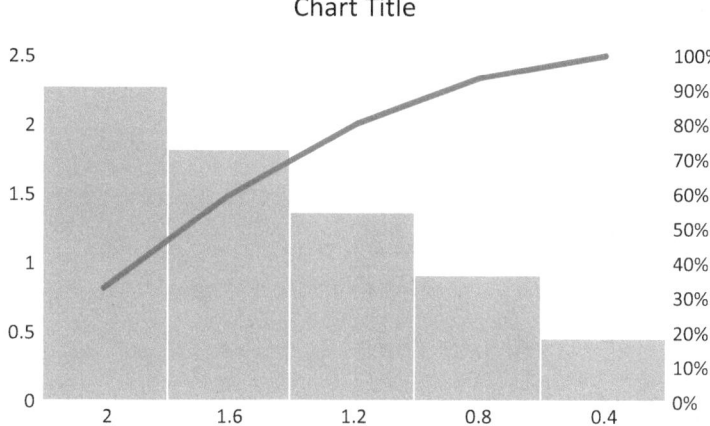

Chart Title

Figure 31: Graph between pitch and surface finish.

The expected outcome for the graduate is to impart knowledge and computation skill for side rake angle.

Learning problem 37

Turning a mild, steel rod at feed of 0.24 mm/rev by a carbide tool having orthogonal rake angle of 30°, the chip thickness was found to be equal to 0.60 mm. Determine and calculate the best chip reduction coefficient.

Bloom's taxonomy cognitive level ACTION VERB: determine

Expected outcome: knowledge (psychomotor domain) pertaining to skill development.

Learning solution 37

$$r = \frac{a_2}{a_1}; \; a_2 = 0.60 \text{ mm and } a_1 = S_0(\sin \varnothing); \; S_0 = 0.24 \text{ mm/rev}$$

"$\varnothing$" is variable

$$r = \frac{a_2}{S_0 \sin \varnothing} = \frac{0.60}{0.24, \; \sin(30°)} = 4.2$$

Table 39: Values of shear angle and chip reduction coefficient.

Ø (degree)	R (chip reduction coefficient)
6	19.13
12	9.619
18	6.477
24	4.917
30	4.2

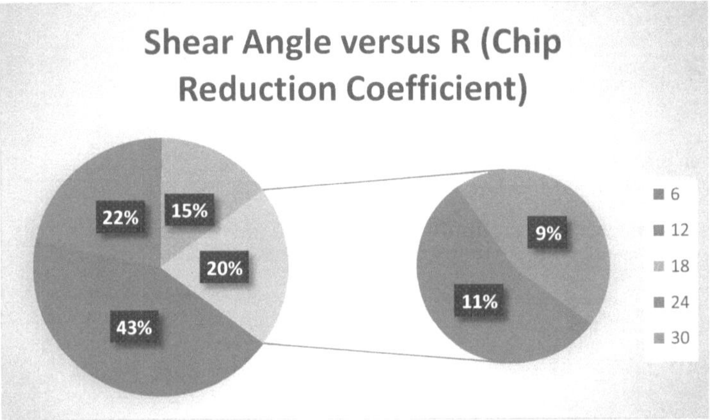

Figure 32: Graph between shear angle and chip reduction coefficient.

The expected outcome for the graduate is to impart knowledge and computation skill for chip reduction coefficient.

Learning problem 38

Tool life is represented by Taylor's tool life equation universally. Calculate and determine the tool life for the given specified conditions:

Condition A. Reduction in tool life from 20 to 12 min as the cutting velocity increased from 200 to 250 m/min. Check for 300 m/min.

Bloom's taxonomy cognitive level ACTION VERB: determine

Expected outcome: knowledge (psychomotor domain) pertaining to skill development.

Learning solution 38

Taylor's tool life eq^n $VT^n = C \rightarrow \left(\dfrac{T_2}{T_1}\right)^n = \left(\dfrac{V_1}{V_2}\right)$

$$\rightarrow \left(\dfrac{12}{20}\right)^n = \left(\dfrac{200}{250}\right) \rightarrow n = 0.35$$

$$\rightarrow T_3 = T_1 \left(\dfrac{V_1}{V_2}\right)^{1/n}$$

$$\rightarrow T_3 = 24 \left(\dfrac{200}{300}\right)^{1/0.55} = 11.48 \; min$$

"Tool life" is variable (T_3).

Table 40: Values of tool life and cutting velocity.

Tool life (min)	Cutting velocity (m/min)
260	14.89
270	13.91
280	13.02
290	12.21
300	11.48

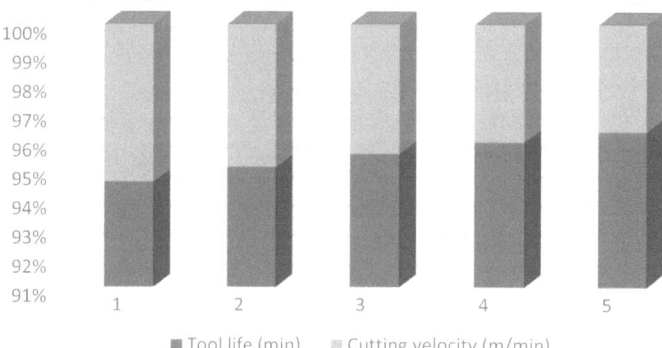

Tool Life versus Cutting Velocity

■ Tool life (min) ■ Cutting velocity (m/min)

Figure 33: Graph between tool life and cutting velocity.

The expected outcome for the graduate is to impart knowledge and computation skill for tool life.

Learning problem 39

Determine and calculate the machining time to reduce the diameter of a rod from 200 to 195 mm over length of 200 mm at cutting velocity of 220 m/min and feed of 0.2 mm/rev.

Bloom's taxonomy cognitive level ACTION VERB: determine and calculate

Expected outcome: knowledge (psychomotor domain) pertaining to skill development.

Learning solution 39

Actual machining time, $T_c = \dfrac{\pi\, D(L_w + A + 0)}{1,000 V_c S_o}$

$$D = 220\,\text{mm},\ L_w = 200\,\text{mm}$$

"V_c" is variable

$$T_c = \frac{\pi(220)(200 + 5 + 5)}{1,000\ (220 \times 0.2)} = 3\,\text{min}$$

Table 41: Values of cutting speed and machining time.

V_c (m/min)	T_c (min)
180	3.66
190	3.47
200	3.29
210	3.14
220	3.0

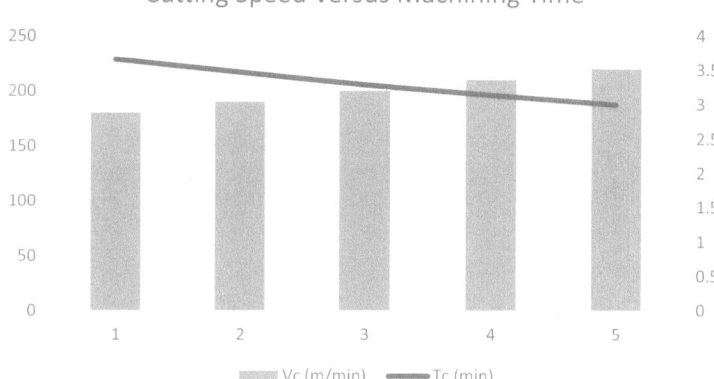

Figure 34: Graph between cutting speed and machining time.

The expected outcome for the graduate is to impart knowledge and computation skill for cutting speed versus machining time.

Learning problem 40

Calculate and determine the maximum total power during the turning operation of Al-6063 alloy for the following input variables:
a) Cutting force, $Z = 710$ N
b) Cutting velocity = 250 m/min

Bloom's taxonomy cognitive level ACTION VERB: determine

Expected outcome: knowledge (psychomotor domain) pertaining to skill development.

Learning solution 40

$$\text{Max total power, } V_t = \frac{[(P_z V_c + P_x V_f)\max + V_f + V_i]f_0 f_t}{n_e n_m}$$

$$V_c = [710 \times 250] = 3.4 \text{ kW}$$

$$V_n = (P_z V_c + P_x V_t) = 3.2 + 0.1 + 3.2 = 3.52 \text{ kW}$$

$$V_t = \frac{(3.52 + 0.15 \times 3.4 + 0.1 + 3.4)1.25 \times 1.2}{0.95 \times 0.9}$$

$$= 13.2 \text{ kW}$$

Table 42: Values of cutting force
and cutting velocity.

V_c	V_t (kW)
200	6.27
210	6.03
220	6.95
230	7.27
240	7.58

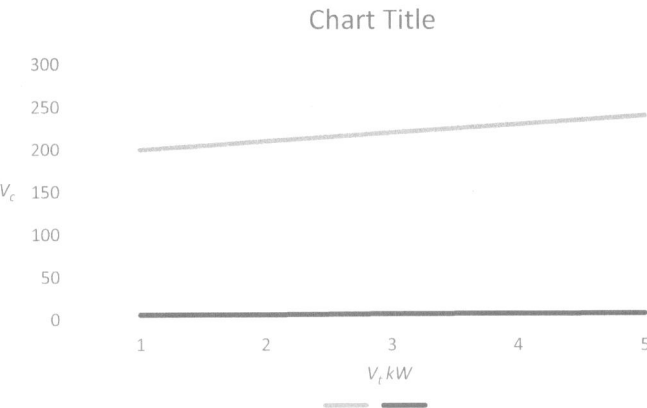

Figure 35: Graph between cutting force and cutting velocity.

The expected outcome for the graduate is to impart knowledge and computation skill maximum power obtained during turning operation.

Chapter 3
Metal cutting optimization

Learning problem 41

Shaper machine tool is widely used for plate to improve the surface quality, which reciprocates in forward and return stroke, with actual cutting in forward stroke only and return stroke is an ideal stroke. Calculate and determine the stroke length for machining cast iron if a slide is reciprocated by a crank of radius 90 mm and a connecting rod length of 180 mm.

Bloom's taxonomy cognitive level ACTION VERB: determine

Expected outcome: knowledge (psychomotor domain) pertaining to skill development.

Learning solution 41

Stroke length; $\qquad S_t = 2 \times$ crank radius
$\qquad$ Quick return ratio $= 1$
$\qquad$ "Crank radius" is variable.

Table 43: Values of crank radius and stroke length.

Crank radius	Stroke length
60	120
70	140
80	160
90	180
100	200

https://doi.org/10.1515/9783110676662-003

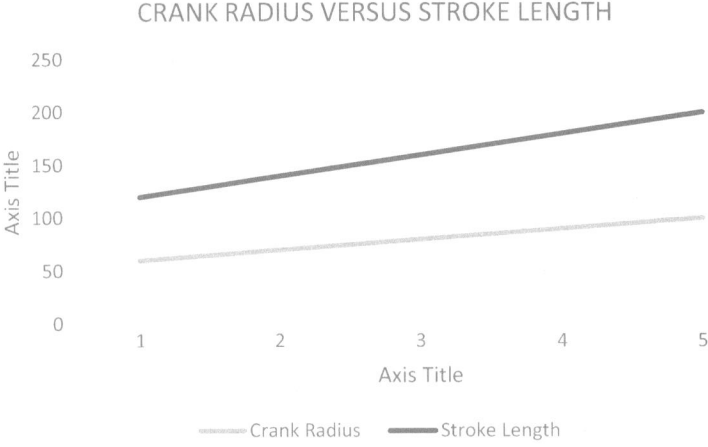

Figure 36: Graph between crank radius and stroke length.

The expected outcome for the graduate is to impart knowledge and computation skill for shaping the engineering material.

Learning problem 42

A center lathe having 14 spindle speeds is recommended for machining Al-6063 alloy (diameter: 50–100 mm) at cutting velocity in between 40 and 210 m/min. Calculate and determine the lowest and the highest spindle speeds of that lathe machine tool utilized for metal cutting operation.

Bloom's taxonomy cognitive level ACTION VERB: determine

Expected outcome: knowledge (psychomotor domain) pertaining to skill development.

Learning solution 42

$$\text{Spindle speed: } N_L = \frac{1,000 \times V_{\min}}{\pi D_{\max}}$$

$$\text{Spindle speed: } N_H = \frac{1,000 \times V_{\max}}{\pi D_{\min}}$$

$$\varnothing \text{ (shear angle)} = \left(\frac{N_H}{N_L}\right)^{1/(z-1)}$$

Table 44: Values of cutting
velocity and shear angle.

Cutting velocity	Ø (shear angle)
160	1.228
170	1.235
180	1.242
190	1.247
200	1.253

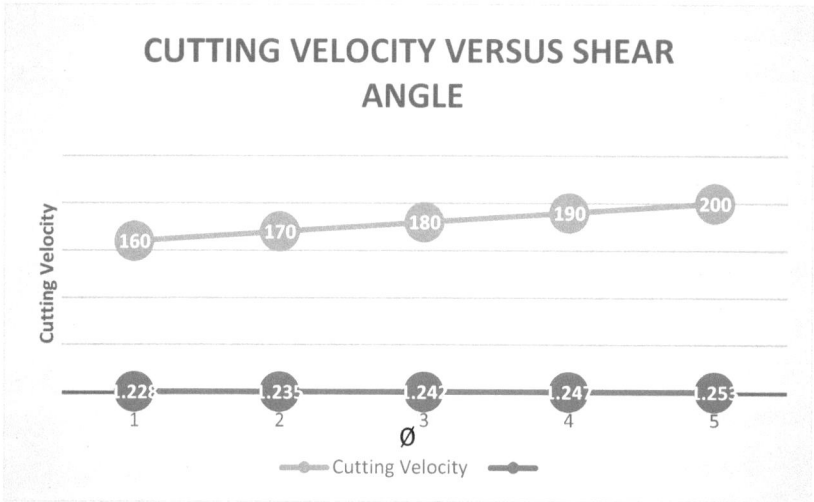

Figure 37: Graph between cutting velocity and shear angle.

The expected outcome for the graduate is to impart knowledge and computation skill for cutting velocity versus shear angle.

Learning problem 43

Calculate and determine the speed of the gear blank on different cutting velocities **for** blank having 42 teeth to be machined using HSS hob cutter (diameter 76 mm) at a cutting velocity of 44 m/min.

Bloom's taxonomy cognitive level ACTION VERB: determine

Expected outcome: knowledge (psychomotor domain) pertaining to skill development.

Learning solution 43

$$\text{Speed, } N_h = \frac{1,000 \times V_c}{\pi D_h} = \frac{1,000 \times 42}{\pi (76)} = 191.4 \text{ RPM}$$

Speed of gear blank, $N_g = 191.4 \times \dfrac{1}{40} = 4.54$ RPM

Table 45: Values of gear blank and cutting velocity.

V_c	N_g
28	3.18
32	3.64
36	4.10
40	4.54
44	5.00

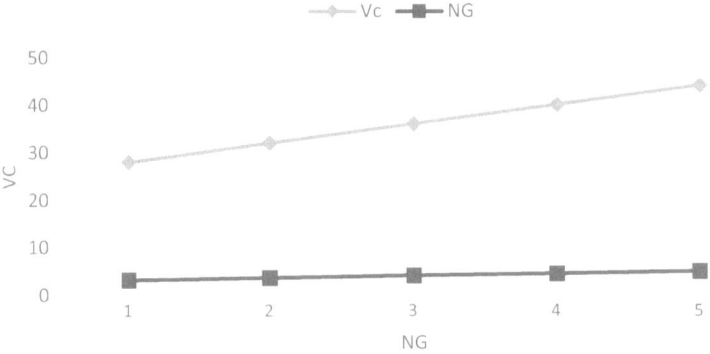

GEAR BLANK SPEED VERSUS CUTTING VELOCITY

Figure 38: Graph between gear blank speed and cutting velocity.

The expected outcome for the graduate is to impart knowledge and computation skill for gear blank speed calculation versus cutting velocity.

Learning problem 44

Calculate and determine the coefficient of friction (M_a) during metal cutting, namely, turning operation on Al-6064 alloy with cutting tool having $r_0 = 0°$ and $\varnothing = 75°$, the magnitudes of the cutting force components P_z and P_x are 710 N and 310 N, respectively.

Bloom's taxonomy cognitive level ACTION VERB: determine

Expected outcome: knowledge (psychomotor domain) pertaining to skill development to machine a taper cut on cylindrical shaft

Learning solution 44

$$\varnothing = 75;\ P_x = P_{xy} \sin \varnothing$$

$$\text{When } r_0 = 0°,\ F = P_{xy} = 410 \text{ N}$$

Normal force, $$N = P_z = 710 \text{ N}$$

$$M_s = F/N = \frac{310}{810} = 0.426$$

Table 46: Values of cutting force and coefficient of friction.

P_z (N)	M_s
600	0.67
650	0.615
700	0.57
750	0.53
5 800	0.50

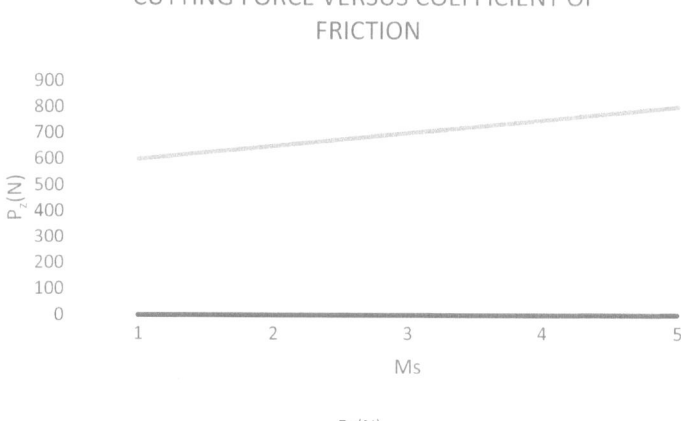

CUTTING FORCE VERSUS COEFFICIENT OF FRICTION

Figure 39: Graph between cutting force and coefficient of friction.

The expected outcome for the graduate is to impart knowledge and computation skill for the coefficient of friction during turning process.

Learning problem 45

Orthogonal cutting operation is to be carried out on Al-6065 alloy. Calculate and determine the coefficient of friction during orthogonal machining operation with the help of the following variables:

$$t_1 = 0.25 \, \text{mm}, \quad \alpha = 0°, \quad F_C = 900\text{N}, \quad F_T = 475\text{N}$$

Bloom's taxonomy cognitive level ACTION VERB: determine and calculate

Expected outcome: knowledge (psychomotor domain) pertaining to skill development.

Learning solution 45

$$\text{Coefficient of friction, } \mu = \frac{F_c \sin \alpha + F_T \cos \alpha}{F_c \cos \alpha - F_T \sin \alpha}$$

$$\mu = 475/900 = 0.34$$

Table 47: Values of rake angle and coefficient of friction.

α (degree)	μ
0	0.5
2	0.544
4	0.59
6	0.638
8	0.689

RAKE ANGLE VERSUS COEFFICIENT OF FRICTION

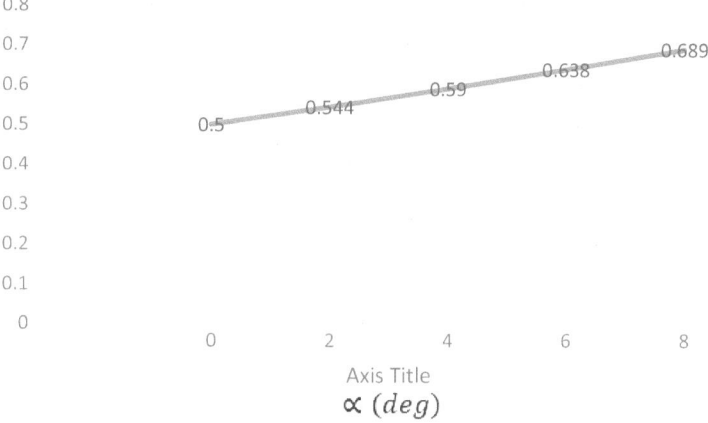

Figure 40: Graph between rake angle and coefficient of friction.

The expected outcome for the graduate is to impart knowledge and computation skill for calculation of coefficient of friction.

Learning problem 46

Steel-graded cylindrical part needs to be machined using HSS tool to reduce the diameter of a steel from 120 to 100 mm area with a length of 200 mm by turning at a cutting velocity (V_c) of 160 m/min. Calculate and determine the machining time required for the change in depth of cut for the given conditions.

Bloom's taxonomy cognitive level ACTION VERB: determine and compute

Expected outcome: knowledge (psychomotor domain) pertaining to skill development.

Learning solution 46

$$\text{Machining time, } T_c = \frac{\pi DL(D_1 - D_2)}{1,000V_cS_o(2t)}$$

$$= \frac{\pi(100)(200)(120 - 100)}{1,000 \times 160 \times 0.25 \times 2}$$

$$= 6.25 \text{ min}$$

Table 48: Values of depth of cut and machining time.

Depth of cut "C" (mm)	M/C time (T_c) (min)
1.2	26.18
1.4	22.44
1.6	19.63
1.8	17.45
5	15.72

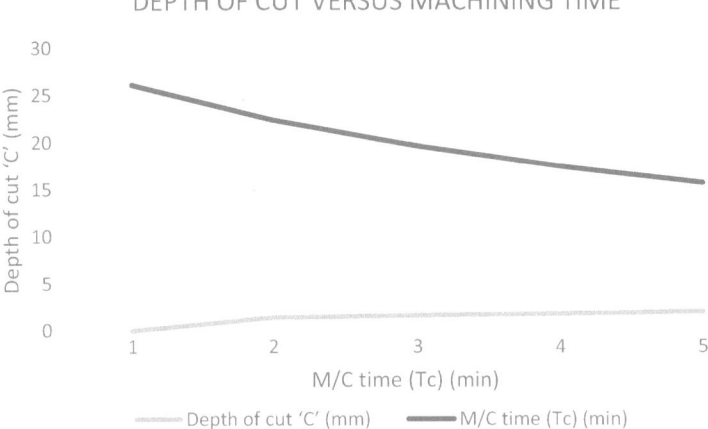

Figure 41: Graph between depth of cut and machining time.

The expected outcome for the graduate is to impart knowledge and computation skill for depth of cut and machining time.

Learning problem 47

Industries have a major challenge to produce large number of products with least time and better quality. Small-scale industries and large-scale industries evaluate the time and life of the parts manufactured. Calculate and determine the machining time using the following input variables, namely, in a batch production of mild steel shafts by machining:

Idle time per piece is 12 min (T_c)

Machining time per piece is 10 min (T_c)

Life of each tool tip is 40 min (TL)

Bloom's taxonomy cognitive level ACTION VERB: determine

Expected outcome: knowledge (psychomotor domain) pertaining to skill development.

Learning solution 47

$$\text{Total machining time, } T_t = T_i + T_c + \frac{T_c}{T_L}(TCT)$$

Table 49: Values of idle time and machining time.

T_c(min)	Total time (T_t) (min)
4	26
8	32
12	38
16	44
20	50

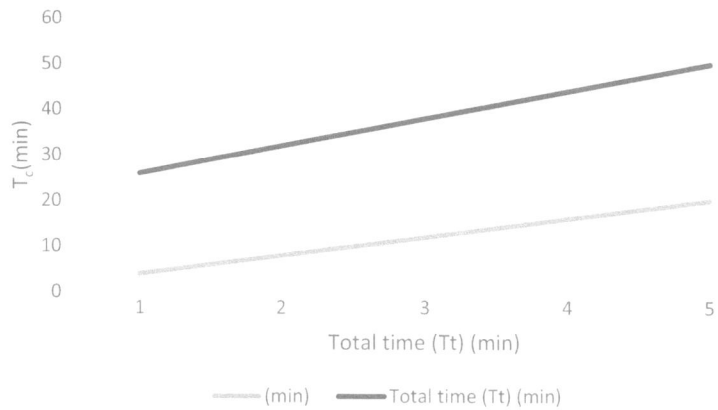

IDLE TIME VERSUS MACHINING TIME

Figure 42: Graph between idle time and machining time.

The expected outcome for the graduate is to impart knowledge and computation skill for machining time.

Learning problem 48

Metal cutting processes are the evidence of vibrations, chatter, and forces due to the contact of cutting tool with the workpiece. Calculate and determine the orthogonal component of cutting force in turning a die steel at a feed of 0.5 mm/rev and depth of cut of 3 mm.

Learning solution 48

$$P_z = t \times S_0 \times C_s (r - \tan r_0 + 1)$$

Table 50: Values of depth of cut and orthogonal force.

Depth of cut (t) (mm)	P_z (N)
0.8	192
1	240
1.5	360
2	480
2.5	600

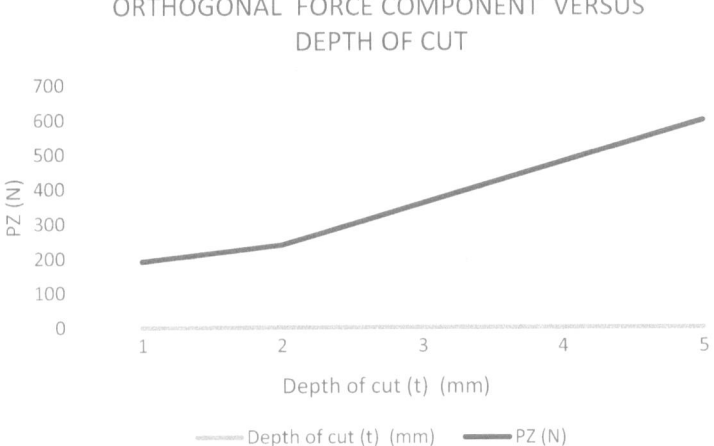

Figure 43: Graph between orthogonal force component and depth of cut.

The expected outcome for the graduate is to impart knowledge and computation skill for cutting force components.

Learning problem 49

Feed rate is provided to the rotary specimen by the cutting tool in the direction longitudinal to the axis of the work specimen. Calculate and determine the Al-6063 roughness value at feed of 0.6 mm/rev which will be the surface roughness if the tool's cutting angles ($\varnothing$ and $\varnothing_1$) are 40° and 20°, respectively.

Bloom's taxonomy cognitive level ACTION VERB: determine and calculate

Expected outcome: knowledge (psychomotor domain) pertaining to skill development.

Learning solution 49

$$\text{Surface roughness, } h_{max} = \frac{S_0}{\cot\varnothing + \cot\varnothing_1}$$

$$= \frac{0.6}{\cot 40° + \cot 20°}$$

$$= 0.0928 \text{ mm}$$

$$= 92.8 \text{ mm}$$

Let $\varnothing_1$ be the variable.

Table 51: Values of surface roughness and shear angle.

$\varnothing_1$ (degree)	h_{max} (mm)
11	69.9
12	75.72
13	81.48
14	87.18
15	92.8

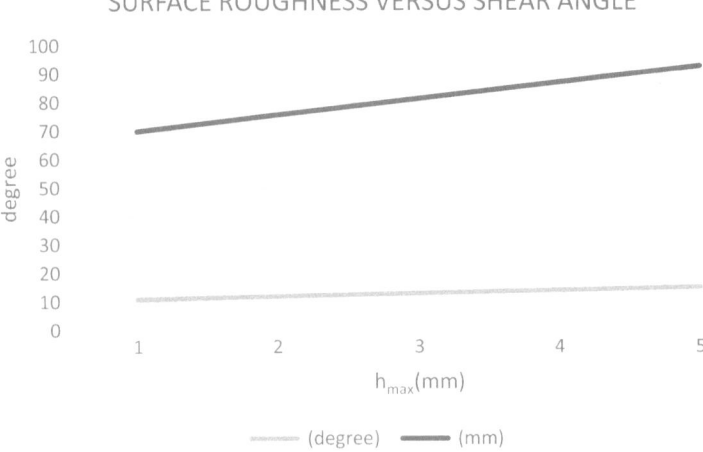

Figure 44: Graph between surface roughness and shear angle.

The expected outcome for the graduate is to impart knowledge and computation skill for testing of surface roughness.

Learning problem 50

The side rake angle in the HSS cutting tool avoids rubbing and provides strength to the cutting tool. Calculate and determine the value of the side rake angle (r_x). The geometries of a single point cutting tool are specified as 0°, 10°, 8°, 6°, 15°, 60°, 0 (mm).

Bloom's taxonomy cognitive level ACTION VERB: determine

Expected outcome: knowledge (psychomotor domain) pertaining to skill development.

Learning solution 50

Side rake angle (r_x)

$$\tan r_x = \tan r_x \cdot \sin \varnothing - \tan \lambda \cos \varnothing$$

$$\therefore \varnothing = \tan(10) \cdot \sin(60) - \tan(0) \cdot \cos(60)$$

Table 52: Values of side rake
angle and shear angle.

Ø (degree)	r_x (degree)
6	5.2
7	6.07
8	6.93
9	7.81
10	8.68

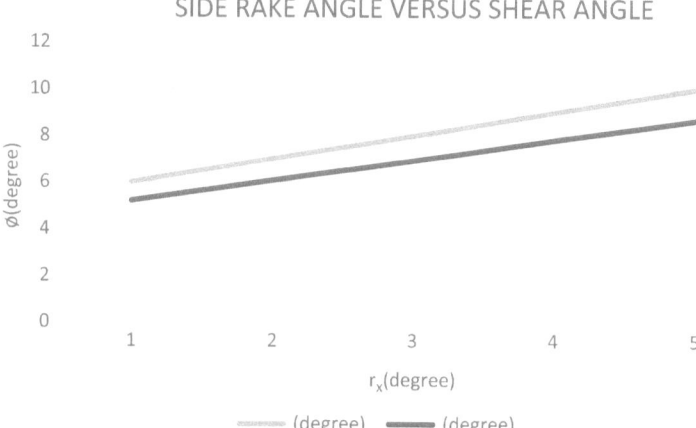

Figure 45: Graph between side rake angle and shear angle.

    The expected outcome for the graduate is to impart knowledge and computation skill for the side rake angle of the cutting tool.

Chapter 4
Metal cutting performance responses

Learning problem 51

Metal cutting of steel grade (low carbon steel) is turned and taper turned using the HSS cutting tool. Calculate and determine the inclination angle (r_0) of the tool in the ASA system as 20°, −20°, 8°, 6°, 15°, 30°, 0 (mm).

Bloom's taxonomy cognitive level ACTION VERB: determine

Expected outcome: knowledge (psychomotor domain) pertaining to skill development.

Learning solution 51

Inclination angle (λ)

$$\tan \lambda = \tan r_y \sin \phi - \tan r_x \cos \phi$$

$$r_x = 20°, \ r_y = -20°, \ \phi = 90 - 30 = 60°$$

Let "ϕ" be the variable

$$\tan \lambda = - \tan(2) \cos(60^0) + \tan(-20^0) \sin 60^0$$

$$\therefore \tan \lambda = 14.54$$

Table 53: Values of inclination angle and shear angle.

Ø (degree)	λ (degree)
20	−12.73
30	−13.95
40	−13.95
50	−12.73
60	−13.54

https://doi.org/10.1515/9783110676662-004

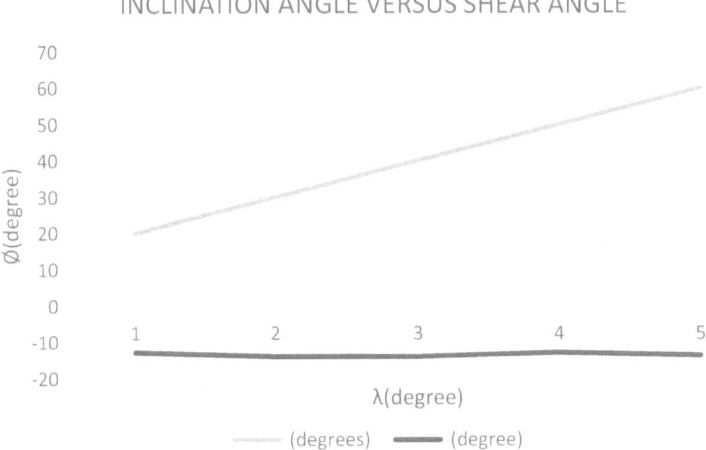

Figure 46: Graph between inclination angle and shear angle.

The expected outcome for the graduate is to impart knowledge and computation skill for the inclination angle.

Learning problem 52

Clearance angle is provided to the HSS cutting tool as one of the tool signatures. Calculate and determine the values of clearance angle ($\propto_m$), where geometry is specified in ORS as –10°, 10°, 8°, 6°, 75°, 0 (mm).

Bloom's taxonomy cognitive level ACTION VERB: determine and calculate

Expected outcome: knowledge (psychomotor domain) pertaining to skill development.

Learning solution 52

Clearance angle ($\propto_m$)

$$\cot \propto_m = \sqrt{\cot^2 \propto_0 + \tan^2 \lambda}$$

$$\therefore \propto_m = \cot^{-1} \sqrt{\cot^2 8 + \tan^2(-10°)}$$

Table 54: Values of clearance angle and ORS.

λ (degree)	α_m (degree)
−10°	7.99
−30°	7.98
−40°	7.94
−50°	7.89
−60°	7.77

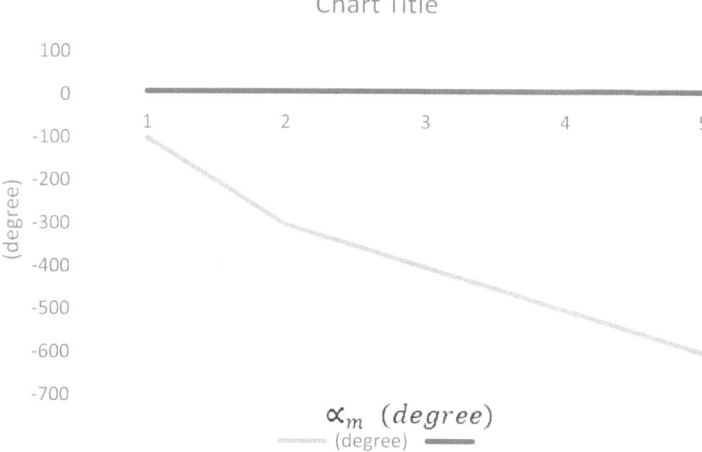

Figure 47: Graph between clearance angle and ORS.

The expected outcome for the graduate is to impart knowledge and computation skill for clearance angle performance for the turning operation.

Learning problem 53

Calculate and determine the axial rake ($Y_x\, D_i$) angle at a radial distance of 4 mm of a 22 mm diameter twist drill with helix angle (θ) is 20.5°.

Bloom's taxonomy cognitive level ACTION VERB: determine and calculate

Expected outcome: knowledge (psychomotor domain) pertaining to skill development.

Learning solution 53

$$\tan Y_{Di} = \left(\frac{Y_i}{Y}\right) \tan \theta$$

$$\tan Y_x D_i = \frac{r_i}{r} \tan \theta$$

Table 55: Values of radial distance and axial angle.

r_i (mm)	$Y_x D_i$ (*degree*)
1	3.57
2	7.12
3	10.61
4	14
5	17.35

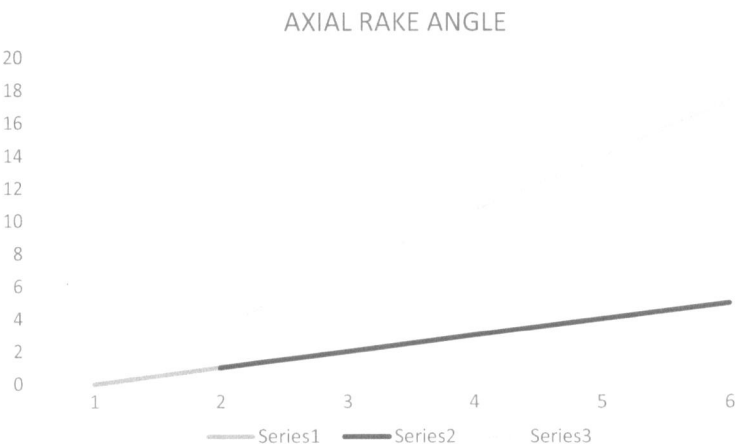

Figure 48: Graph between radial distance and axial rake angle.

The expected outcome for the graduate is to impart knowledge and computation skill for the axial rake angle.

Learning problem 54

Calculate and determine the side rake angle for the cutting tool having thread pitch of 5.0 mm that cuts on a 11.5 mm diameter rod by the HSS cutting tool with geometries 0°, 0°, 8°, 10°, 30°, 30°, 0 (mm).

Bloom's taxonomy cognitive level ACTION VERB: determine

Expected outcome: knowledge (psychomotor domain) pertaining to skill development.

Learning solution 54

$$Y_{XW} = Y_{XDi} + r_i$$

$$\tan u_i = \frac{S_0}{2\pi r}$$

Here, S_0 = pitch = 5 mm, r = 5.5 mm

$$\therefore u_i = \tan^{-1}\left(\frac{x}{x\pi 8}\right)$$

Table 56: Values of thread pitch and side rake angle.

π mm	Y_{XW} (degree)
8	2.27
7	2.60
6	3.03
5	3.64
4	4.55

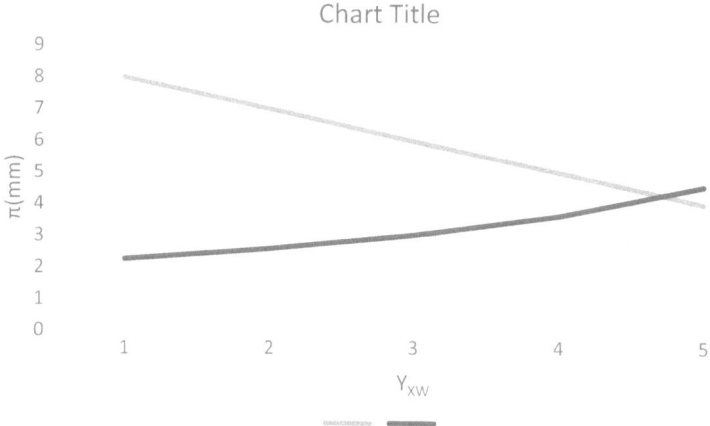

Figure 49: Graph between thread pitch and side rake angle.

The expected outcome for the graduate is to impart knowledge and computation skill for the side rake angle.

Learning problem 55

Determine and calculate the friction angle with variation of shear angle for a die steel rod specimen at feed 0.35 mm/rev by a carbide tool having orthogonal rake angle of 20° and principal cutting edge angle of 30°.

Bloom's taxonomy cognitive level ACTION VERB: determine and calculate

Expected outcome: knowledge (psychomotor domain) pertaining to skill development.

Learning solution 55

Let r = chip reduction coefficient

$$\therefore r = \frac{a_2}{a_1} = \frac{0.48}{0.35 \sin 30°} = 3.84$$

$$r = S_0 \sin \varnothing \text{ and } \varnothing = 30°$$

$$\tan \beta = \frac{\cos Y_0}{r - \sin Y_0}, \text{ where } Y_0 = 20 \text{ orthogonal rake angle}$$

$$\therefore \beta = \tan^{-1}\left(\frac{\cos 20}{3.84 - \sin 10}\right) = 15.03$$

Table 57: Values of rake angle and cutting angle.

$\varnothing$ (degree)	β (degree)
6	2.97
12	5.95
18	8.887
24	11.77
30	14.4

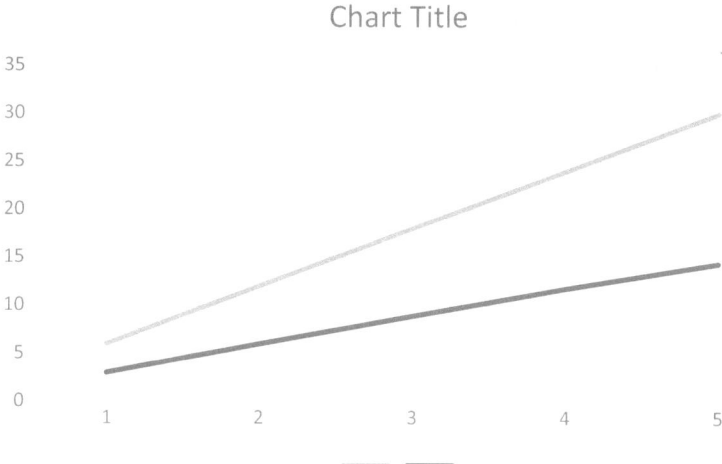

Figure 50: Graph between rake angle and cutting angle.

The expected outcome for the graduate is to impart knowledge and computation friction angle.

Learning problem 56

Determine and calculate the temperature of the HSS cutting tool during machining of die steel on the lathe machine tool. Under a given condition of plain turning of a mild steel rod by an HSS tool, the average cutting year temperature was measured to be around 565 °C.

Bloom's taxonomy cognitive level ACTION VERB: determine and calculate

Expected outcome: knowledge (psychomotor domain) pertaining to skill development.

Learning solution 56

$$\theta_{avg} \propto (a_1)^{2.4}, \ \theta_{vag} = 565 \, C$$

$$avg_2 = \left(\frac{\sin \theta}{\sin 90}\right)^{0.24} \times 565°$$

$$avg_2 = \left[\frac{\sin 30}{\sin 90}\right] \times 565 = 502.2 \, C$$

Table 58: Values of average cutting temperature and ∅.

∅ (degree)	avg$_2$ (C)
30	508.05
40	539.62
50	562.82
60	579.64
70	591.11

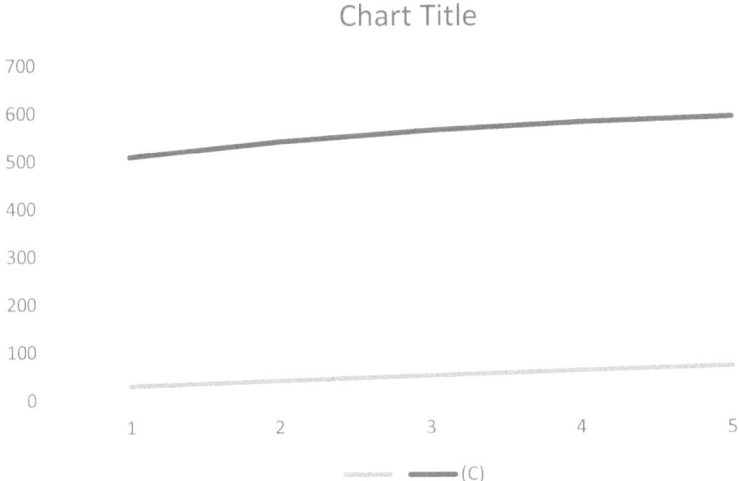

Figure 51: Graph between average cutting temperature and ∅.

The expected outcome for the graduate is to impart knowledge and computation skill for temperature measurement of cutting tool during the metal cutting process.

Learning problem 57

Calculate and determine chip tool contact length for the machining of die-steel with following input variables, namely, 10° orthogonal rake angle/75° principal cutting edge (∅)/feed, 0.32 mm/rev.

Bloom's taxonomy cognitive level ACTION VERB: calculate and determine

Expected outcome: knowledge (psychomotor domain) pertaining to skill development.

Learning solution 57

Given $a_2 = 0.70$ mm, $S_0 = 0.32$ mm/rev, $\varnothing = 75°$, $Y_0 = 10°$

Let the entire chip be in plastic contact. Now,

$$C_p = a_1\left[1 + \tan\left(\beta_0 Y_1\right)\right]$$

$$\tan\beta = \frac{\cos Y_0}{r - \sin Y_0}, \quad 2.26$$

$$\therefore \tan\beta = \frac{\cos 10}{2.26 - 0.966} = 0.56$$

$$\therefore \beta = 37.27°$$

$$\therefore C_p = a_2\left[1 + \tan\left(37.27° - 10°\right)\right]$$

$$= 1.02 \text{ mm}$$

Table 59: Values of orthogonal rake angle and C_p.

Y (degree)	C_p (mm)
10	0.8
9	0.816
8	0.827
7	0.837
6	0.847

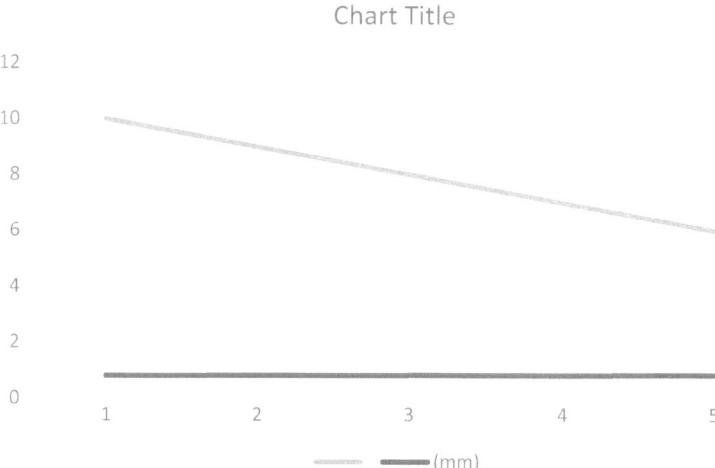

Chart Title

Figure 52: Graph between orthogonal rake angle and C_p.

The expected outcome for the graduate is to impart knowledge and computation skill for the chip tool contact length.

Learning problem 58

Calculate and determine the machining power for the metal cutting turning process for Al-6063 alloy having diameter of 210 mm and velocity of 120 m/min with feed of 0.2 mm/rev.

Bloom's taxonomy cognitive level ACTION VERB: calculate and determine

Expected outcome: knowledge (psychomotor domain) pertaining to skill development.

Learning solution 58

$$S_0 = \text{feed} = 0.2\,\text{mm/rev}, \quad d = \text{diameter} = 210 \text{ mm}$$

$$V_c = \text{cutting velocity} = 120 \text{ m/min}$$

$$n = \text{spindle speed.}$$

$$V_c = \frac{\pi d n}{1,000} = 120\,\text{m/min} = Z_x S^{-1}$$

$$\therefore n = \frac{1,000 \times 120}{\pi \times 200} = 191 \text{ RPM}$$

$$\therefore \text{Total} \frac{M}{C} \text{power,} \quad E_T = P_Z V_C + P_\pi V_{\text{feed}}$$

$$\therefore E_T = 800 \times 2 + 450\,N \times \frac{0.038}{60}$$

$$= 1,600 + 0.2865$$

$$= 1,600 \text{ W}$$

Table 60: Values of cutting velocity and power.

V_c m/s	E_T (W)
1.2	960
1.4	1,120
1.6	1,280
1.8	1,440
2	1,600

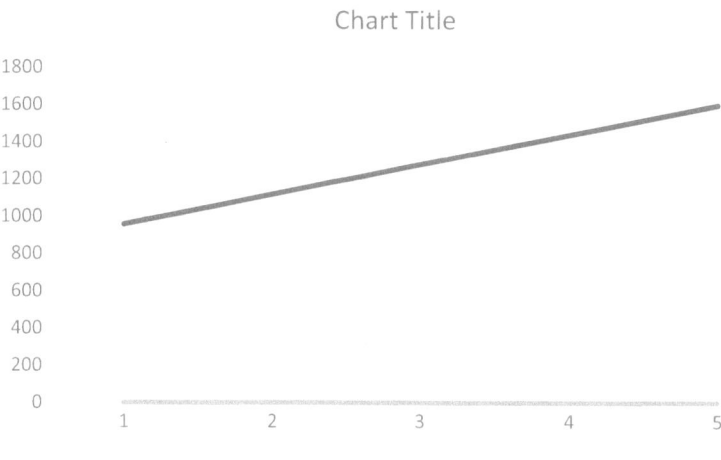

Chart Title

Figure 53: Graph between cutting velocity and power.

The expected outcome for the graduate is to impart knowledge and computation skill for the machining power.

Learning problem 59

Calculate and determine the average chip tool interface temperature during the machining of stainless steel rods having yield shear strength of 420 and 200 MPa at cutting velocities of 120 and 220 m/min.

Bloom's taxonomy cognitive level ACTION VERB: determine

Expected outcome: knowledge (psychomotor domain) pertaining to skill development.

Learning solution 59

$$\frac{\theta_{ib}}{\theta_{ia}} = \left(\frac{Z_{sb}}{Z_{sa}}\right)\left(\frac{V_{cb}}{V_{ca}}\right) = \left(\frac{220}{420}\right)\left(\frac{220}{120}\right) = 0.85$$

Table 61: Values of cutting velocity and ratio.

V_{cb} (m/min)	Ratio
50	0.25
100	0.5
150	0.75
175	0.875
200	1

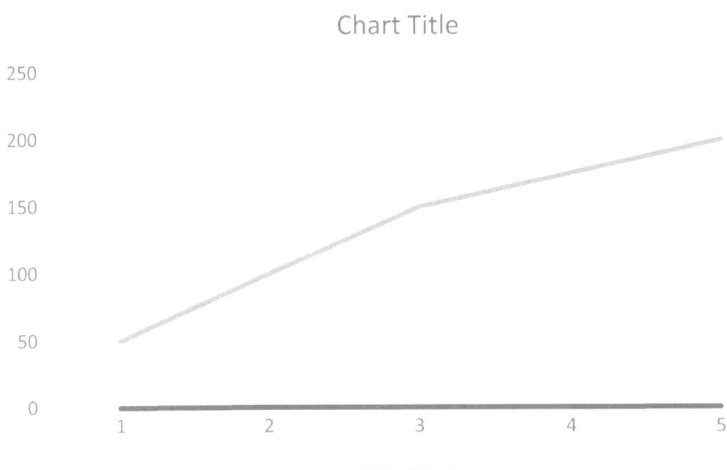

Figure 54: Graph between cutting velocity and ratio.

The expected outcome for the graduate is to impart knowledge and computation skill for the measurement of chip tool interface temperature.

Learning problem 60

Determine and calculate the machining time required to reduce the diameter of an Al-6061 cylindrical rod from 220 to 195 mm over a length of 300 mm.

Bloom's taxonomy cognitive level ACTION VERB: determine and calculate

Expected outcome: knowledge (psychomotor domain) pertaining to skill development.

Learning solution 60

Formula for machining time, $T_c = \dfrac{\pi D(L_w + A + 0)}{1000 V_c S_o}$

$$V_c = 220 \text{ m/min}, \ S_o = 0.2 \text{ mm/rev}$$

$$\therefore T_c = \frac{\pi\,(220)(300 + 5 + 5)}{1,000 \times (220 \times 0.2)} = 4.32 \text{ min}$$

Table 62: Values of machining time and diameter.

Diameter, D (mm)	T_c (min)
50	0.75
100	1.5
150	2.25
175	2.62
200	3

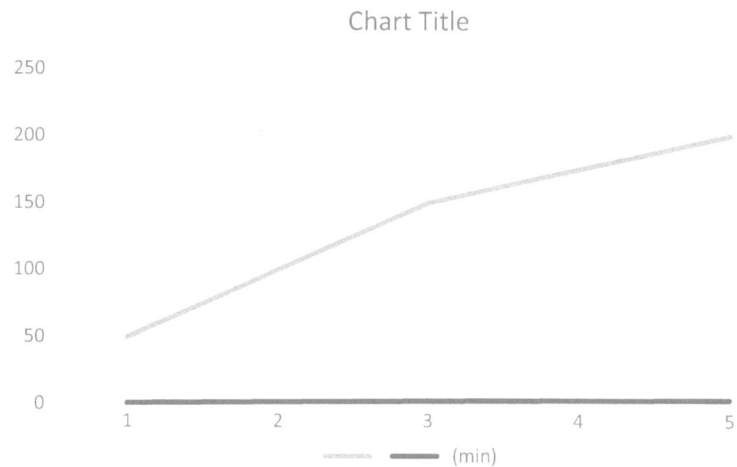

Figure 55: Graph between machining time and diameter.

The expected outcome for the graduate is to impart knowledge and computation skill for the machining time calculation.

Learning problem 61

Grinding is a surface finish process needed to perform on all metallic plates and cylindrical parts after primary machining process, namely, tuning, milling, welding, and casting. Calculate the metal removal rate for the machining of mild steel rod with input set at depth of cut = 0.03 mm and wheel speed = 100 RPM.

Bloom's taxonomy cognitive level ACTION VERB: determine

Expected outcome: knowledge (psychomotor domain) pertaining to skill development.

Learning solution 61

Material removal rate

$$Z = V_t \times 1,000 \times f \times d \ \ (\text{mm}^3/\text{min})$$
$$= 30 \times 1,000 \times 10 \times 0.05$$
$$Z = 15,000 \, \text{mm}^3/\text{min}$$

Table 63: Values of depth of cut and wheel speed.

(V_t)	(Z)
4	1,000
8	2,000
12	3,000
16	4,000
20	5,000

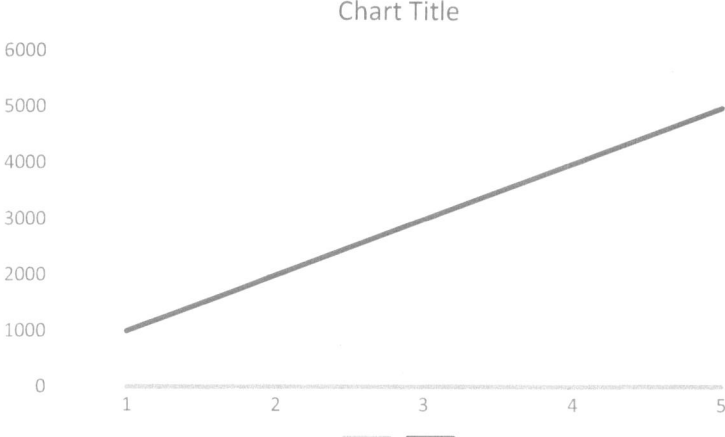

Figure 56: Graph between depth of cut and wheel speed.

The expected outcome for the graduate is to impart knowledge and computation skill for the calculation of metal removal rate.

Learning problem 62

Hard and difficult-to-machine materials are not easy to machine using conventional machining processes. ECM (electrochemical machining) is utilized for the machining of metallic Fe base plate. Calculate and determine the current needed if the atomic weight of iron is 56, valency = 2, and density = 7.8 g/cm³.

Bloom's taxonomy cognitive level ACTION VERB: determine

Expected outcome: knowledge (psychomotor domain) pertaining to skill development.

Learning solution 62

$$\text{Metal removal rate (MRR)} = \frac{EI}{F\int} \left[cm^3/5 \right]$$

$$I = \frac{(MRR)F\int}{E}$$

MRR = 3 cm³/min (MRR is variable).
$F = 26.8 \text{ amp} - hr = 1,008 \text{ amp} - min$

$$E = \frac{N}{n} = \frac{60}{2} = 30$$

$$I = \frac{3 \times 1,608 \times 7.8}{30} = 1,254.24 \text{ A}$$

Table 64: Values of current and MRR.

MRR (cm^3/min)	Current (I)
0.6	268.76
1.2	537.53
1.8	806.29
2.4	1,075.06
3	1,343.8

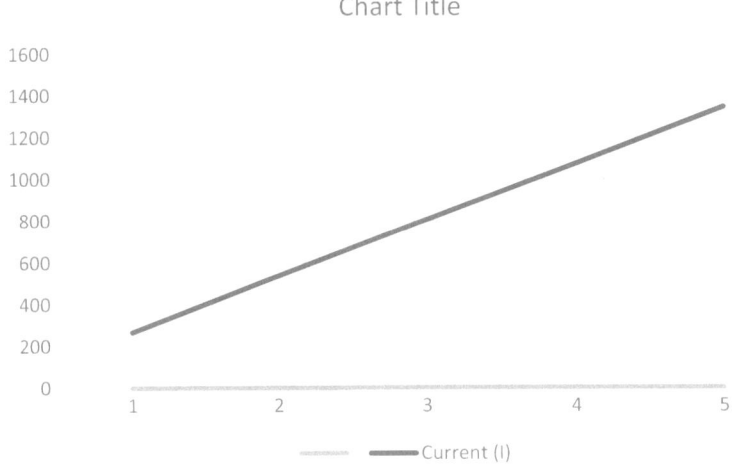

Chart Title

Current (I)

Figure 57: Graph between current and MRR.

The expected outcome for the graduate is to impart knowledge and computation skill for the metal removal rate calculation in the ECM process.

References

Abdullah, Z. T., Sheng, G. S., Yun, S. B. (2018). Conventional milling machine into CNC machine tool remanufacturing, eco-comparison ratio based analysis. *Current Journal of Applied Science and Technology*, *28*(6), 1–18.

Alawad, M. N., Fattah, K. A. (2019). Superior fracture-seal material using crushed date palm seeds for oil and gas well drilling operations. *Journal of King Saud University-Engineering Sciences*, *31*(1), 97–103.

Alphonse, M., Raja, V. B., Gupta, M. (2020). Investigation on tribological behavior during friction drilling process-A review. *Tribology in Industry*, *42*(2), 200.

Antonialli, A. Í. S., Magri, A., Diniz, A. E. (2016). Tool life and tool wear in taper turning of a nickel-based superalloy. *The International Journal of Advanced Manufacturing Technology*, *87*(5), 2023–2032.

Arunkarthikeyan, K., Balamurugan, K. (2021). Experimental studies on deep cryo treated plus tempered tungsten carbide inserts in turning operation. In Arockiarajan, A., Duraiselvam, M., Raju, R. (eds.) *Advances in Industrial Automation and Smart Manufacturing* 313–323. Springer, Singapore. https://doi.org/10.1007/978-981-15-4739-3_26.

Ataallahi, E., Shadizadeh, S. R. (2015). Fuzzy consequence modeling of blowouts in Iranian drilling operations; HSE consideration. *Safety Science*, *77*, 152–159.

Azim, S., Noor, S., Khalid, Q. S., Khan, A. M., Pimenov, D. Y., Ahmad, I., . . . Pruncu, C. I. (2020). Sustainable manufacturing and parametric analysis of mild steel grade 60 by deploying CNC milling machine and Taguchi method. *Metals*, *10*(10), 1303.

Baron, P., Dobránsky, J., Pollák, M., Cmorej, T., Kočiško, M. (2015). Proposal of the knowledge application environment of calculating operational parameters for conventional machining technology. In *Key Engineering Materials, 669*, 95–102. https://doi.org/10.4028/www.scientific.net/kem.669.95

Bilga, P. S., Singh, S., Kumar, R. (2016). Optimization of energy consumption response parameters for turning operation using Taguchi method. *Journal of Cleaner Production*, *137*, 1406–1417.

Boswell, B., Islam, M. N., Davies, I. J., Ginting, Y. R., Ong, A. K. (2017). A review identifying the effectiveness of minimum quantity lubrication (MQL) during conventional machining. *The International Journal of Advanced Manufacturing Technology*, *92*(1), 321–340.

Boyou, N. V., Ismail, I., Sulaiman, W. R. W., Haddad, A. S., Husein, N., Hui, H. T., Nadaraja, K. (2019). Experimental investigation of hole cleaning in directional drilling by using nano-enhanced water-based drilling fluids. *Journal of Petroleum Science and Engineering*, *176*, 220–231.

Budin, S., Maideen, N. C., Sahudin, S. (2019, May). Design and development of stirring tool pin profile for reconfigured milling machine to perform friction stir welding process. In *IOP Conference Series: Materials Science and Engineering* (Vol. 505, No. 1, p. 012089). IOP Publishing.

Chen, L., Wang, X., Zhang, D., Yu, H., Yang, Q., Lou, D., . . . Sheng, K. (2017). Influence of Cooling Oil Temperature on Cutting Stability of Gear Shaping Machine. *Machine Tool & Hydraulics*, 08.

Chen, S. H., Fong, Z. H. (2015). Study on the cutting time of the hypoid gear tooth flank. *Mechanism and Machine Theory*, *84*, 113–124.

Dambhare, S., Deshmukh, S., Borade, A., Digalwar, A., Phate, M. (2015). Sustainability issues in turning process: A study in Indian machining Industry. *Procedia CIRP*, *26*, 379–384.

Denis, A., Jaeger, J., Taboada, H. (2018, August). Progress thread placement for overlapping MPI non-blocking collectives using simultaneous multi-threading. In *European Conference on Parallel Processing* (pp. 123–133). Springer, Cham.

https://doi.org/10.1515/9783110676662-005

Erkorkmaz, K., Katz, A., Hosseinkhani, Y., Plakhotnik, D., Stautner, M., Ismail, F. (2016). Chip geometry and cutting forces in gear shaping. *CIRP Annals*, *65*(1), 133–136.

Eshiet, K. I. I., Sheng, Y. (2018). The performance of stochastic designs in wellbore drilling operations. *Petroleum Science*, *15*(2), 335–365.

Fayzimatov, S. N., Xusanov, Y. Y., Valixonov, D. A. (2021). Optimization conditions of drilling polymeric composite materials. *The American Journal of Engineering and Technology*, *3*(02), 22–30.

García-Nieto, P. J., García-Gonzalo, E., Vilán, J. V., Robleda, A. S. (2016). A new predictive model based on the PSO-optimized support vector machine approach for predicting the milling tool wear from milling runs experimental data. *The International Journal of Advanced Manufacturing Technology*, *86*(1), 769–780.

Gupta, G., Mishra, R. P. (2017). A failure mode effect and criticality analysis of conventional milling machine using fuzzy logic: Case study of RCM. *Quality and Reliability Engineering International*, *33*(2), 347–356.

Gupta, P., Singh, B. (2020). Local mean decomposition and artificial neural network approach to mitigate tool chatter and improve material removal rate in turning operation. *Applied Soft Computing*, *96*, 106714.

Gupta, S., Chadha, V., Sardana, V., Setia, V., Singari, R. M. Experimental Analysis of Surface Roughness in CNC Taper Turning of Aluminum 6061 Using Taguchi Technique.

Hankins, D., Salehi, S., Karbalaei Saleh, F. (2015). An integrated approach for drilling optimization using advanced drilling optimizer. *Journal of Petroleum Engineering, 215,* 2015.

Karuppusamy, S., Kumar, B. S., Kumar, M. R., Guru, K. R., Rameshbabu, A. M. (2021). Analysis of square threading process by using response surface methodology. *Materials Today: Proceedings*, *37*, 3417–3422.

Katz, A., Erkorkmaz, K., Ismail, F. (2018). Virtual model of gear shaping – part ii: Elastic deformations and virtual gear metrology. *Journal of Manufacturing Science and Engineering*, *140*(7), 071008.

Khalilpourazari, S., Khalilpourazary, S. (2017). A lexicographic weighted Tchebycheff approach for multi-constrained multi-objective optimization of the surface grinding process. *Engineering Optimization*, *49*(5), 878–895.

Khoshdarregi, M. R., Altintas, Y. (2015). Generalized modeling of chip geometry and cutting forces in multi-point thread turning. *International Journal of Machine Tools & Manufacture*, *98*, 21–32.

Kirpihnikova, I. M., Makhsumov, I. B., Nosirov, I. S. (2018, October). Electric servo drive control system of milling machine with neural network. In *2018 International Ural Conference on Green Energy (UralCon)* (pp. 223–226). IEEE.

Korotun, M., Denysenko, Y., Ciszak, O., Ivchenko, O. (2021). Improvement of the gear shaping effectiveness for bimetal gears of internal gearing with a friction coating. In: Ivanov, V., Trojanowska, J., Pavlenko, I., Zajac, J., Peraković, D. (eds.) *Advances in design, simulation and manufacturing IV. DSMIE 2021. Lecture Notes in Mechanical Engineering*. Springer, Cham. https://doi.org/10.1007/978-3-030-77719-7_44.

Kumar, C. V., Vardhan, H., Murthy, C. S. (2019). Quantification of rock properties using frequency analysis during diamond core drilling operations. *Journal of the Institution of Engineers (India): Series D*, *100*(1), 67–81.

Lam, J. (2016). *Utilizing Multi-Threading, Parallel Processing, and Memory Management Techniques to Improve Transportation Model Performance* (No. 16–2386).

Li, M., Yu, T., Zhang, R., Yang, L., Ma, Z., Li, B., . . . Zhao, J. (2020). Experimental evaluation of an eco-friendly grinding process combining minimum quantity lubrication and graphene-enhanced plant-oil-based cutting fluid. *Journal of Cleaner Production*, *244*, 118747.

Lorenzon, A. F., Cera, M. C., Beck, A. C. S. (2015). Performance and energy evaluation of different multi-threading interfaces in embedded and general purpose systems. *Journal of Signal Processing Systems*, *80*(3), 295–307.

Mohamad, A., Zain, A. M., Mohd Yusof, N., Najarian, F., Alwee, R., Abdull Hamed, H. N. (2019). Modeling and optimization of machining parameters using regression and Cuckoo search in deep hole drilling process. *Applied Mechanics and Materials, 892,* 177–184. https://doi.org/10.4028/www.scientific.net/amm.892.177

Mohapatra, K. D., Sahoo, S. K. (2015). Experimental investigation of wire EDM parameters for gear cutting process using desirability with PCA. *International Journal for Technological Research in Engineering, 2*(10), 2415–2419.

Okoro, E. E., Alaba, A. O., Sanni, S. E., Ekeinde, E. B., Dosunmu, A. (2019). Development of an automated drilling fluid selection tool using integral geometric parameters for effective drilling operations. *Heliyon, 5*(5), e01713.

Öztürk, B. (2019). Energy consumption model for the pipe threading process using 10 wt.-% Cu and 316L stainless steel powder-reinforced aluminum 6061 fittings. *Materials Testing, 61*(8), 797–805.

Pandivelan, C., Jeevanantham, A. (2015). Formability evaluation of AA 6061 alloy sheets on single point incremental forming using CNC vertical milling machine. *Journal of Materials and Environmental Science, 6*(5), 1343–1353.

Pandiyan, V., Caesarendra, W., Tjahjowidodo, T., Praveen, G. (2017). Predictive modelling and analysis of process parameters on material removal characteristics in abrasive belt grinding process. *Applied Sciences, 7*(4), 363.

Park, J., Law, K. H., Bhinge, R., Biswas, N., Srinivasan, A., Dornfeld, D. A., . . . Rachuri, S. (2015, June). A generalized data-driven energy prediction model with uncertainty for a milling machine tool using Gaussian Process. In *International Manufacturing Science and Engineering Conference* (Vol. 56833, p. V002T05A010). American Society of Mechanical Engineers.

Parthiban, P., Ayyasamy, I. R., Mathiazhagan, I. N. Gear Shaping Attachment in a Shaper Machine.

Patra, K., Jha, A. K., Szalay, T., Ranjan, J., Monostori, L. (2017). Artificial neural network based tool condition monitoring in micro mechanical peck drilling using thrust force signals. *Precision Engineering, 48,* 279–291.

Pedersen, T., Godhavn, J. M., Schubert, J. (2015). Supervisory control for underbalanced drilling operations. *IFAC-PapersOnLine, 48*(6), 120–127.

Pimenov, D. Y., Guzeev, V. I., Krolczyk, G., Mia, M., Wojciechowski, S. (2018). Modeling flatness deviation in face milling considering angular movement of the machine tool system components and tool flank wear. *Precision Engineering, 54,* 327–337.

Piska, M., Sliwkova, P. (2015). A study of cutting and forming threads with coated HSS taps. *Journal of Machine Engineering, 15*(3), 15.

Pollock, J., Stoecker-Sylvia, Z., Veedu, V., Panchal, N., Elshahawi, H. (2018, April). Machine learning for improved directional drilling. In *Offshore Technology Conference.* OnePetro.

Prasad, B. S., Babu, M. P., Reddy, Y. R. (2016). Evaluation of correlation between vibration signal features and three-dimensional finite element simulations to predict cutting tool wear in turning operation. *Proceedings of the Institution of Mechanical Engineers, Part B: Journal of Engineering Manufacture, 230*(2), 203–214.

Raia, M. R., Ciceo, S., Chauvicourt, F., Martis, C. (2020, October). Influence of stator teeth harmonic shaping on the vibration response of an electrical machine. In *2020 International Conference and Exposition on Electrical And Power Engineering (EPE)* (pp. 193–199). IEEE.

Rao, R. V., Rai, D. P., Balic, J. (2016). Surface grinding process optimization using Jaya algorithm. In: Behera, H., Mohapatra, D. (eds.) *Computational intelligence in data mining – volume 2. Advances in intelligent systems and computing,* vol 411. Springer, New Delhi. https://doi.org/10.1007/978-81-322-2731-1_46.

Revuru, R. S., Zhang, J. Z., Posinasetti, N. R., Kidd, T. (2018). Optimization of titanium alloys turning operation in varied cutting fluid conditions with multiple machining performance characteristics. *The International Journal of Advanced Manufacturing Technology, 95*(1), 1451–1463.

Rudrapati, R., Pal, P. K., Bandyopadhyay, A. (2016). Modeling and optimization of machining parameters in cylindrical grinding process. *The International Journal of Advanced Manufacturing Technology, 82*(9–12), 2167–2182.

Sabkhi, N., Pelaingre, C., Barlier, C., Moufki, A., Nouari, M. (2015). Characterization of the cutting forces generated during the gear hobbing process: Spur gear. *Procedia CIRP, 31*, 411–416.

Saikaew, C. (2018). An implementation of measurement system analysis for assessment of machine and part variations in turning operation. *Measurement, 118*, 246–252.

Samyan, Q. W., Sahar, W., Talha, W., Aslam, M., Martinez-Enriquez, A. M. (2015, October). Real time digital image processing using point operations in multithreaded systems. In *2015 Fourteenth Mexican International Conference on Artificial Intelligence (MICAI)* (pp. 52–57). IEEE.

Sangeetha, G., Arun kumar, B., Srinivas, A., Siva Krishna, A., Gobinath, R., Awoyera, P.O. (2020). Optimization of drilling rig hydraulics in drilling operations using soft computing techniques. In: Das, K., Bansal, J., Deep, K., Nagar, A., Pathipooranam, P., Naidu, R. (eds.) *Soft computing for problem solving. Advances in intelligent systems and computing*, vol 1048. Springer, Singapore. https://doi.org/10.1007/978-981-15-0035-0_69

Saravanakumar, A., Dhanabal, S., Jayanand, E., Logeshwaran, P. (2018). Analysis of process parameters in surface grinding process. *Materials Today: Proceedings, 5*(2), 8131–8137.

Selvam, M. D., Senthil, P. (2016). Investigation on the effect of turning operation on surface roughness of hardened C45 carbon steel. *Australian Journal of Mechanical Engineering, 14*(2), 131–137.

Shivakoti, I., Kibria, G., Pradhan, P. M., Pradhan, B. B., Sharma, A. (2019). ANFIS based prediction and parametric analysis during turning operation of stainless steel 202. *Materials and Manufacturing Processes, 34*(1), 112–121.

Shokoohi, Y., Khosrojerdi, E., Shiadhi, B. R. (2015). Machining and ecological effects of a new developed cutting fluid in combination with different cooling techniques on turning operation. *Journal of Cleaner Production, 94*, 330–339.

Sieczkarek, P., Wernicke, S., Gies, S., Tekkaya, A. E., Krebs, E., Wiederkehr, P., . . . Stangier, D. (2017). Improvement strategies for the formfilling in incremental gear forming processes. *Production Engineering, 11*(6), 623–631.

Silva, L. R., Corrêa, E. C., Brandao, J. R., de Avila, R. F. (2020). Environmentally friendly manufacturing: Behavior analysis of minimum quantity of lubricant-MQL in grinding process. *Journal of Cleaner Production, 256*, 103287.

Simionescu, C. S., Debeleac, C. (2015). The influence of the gear cutting dynamics upon the shape and position of the replaceable cutting edges made from sintered metallic carbides. In *Mathematics and Computer in Science and Industry, Varna*. 127–131. ISBN: 978-1-61804-247-7.

Tanvir, M. H., Hussain, A., Rahman, M. M., Ishraq, S., Zishan, K., Rahul, S. K., Habib, M. A. (2020). Multi-objective optimization of turning operation of stainless steel using a hybrid whale optimization algorithm. *Journal of Manufacturing and Materials Processing, 4*(3), 64.

Tao, J., Qin, C., Xiao, D., Shi, H., Ling, X., Li, B., Liu, C. (2020). Timely chatter identification for robotic drilling using a local maximum synchrosqueezing-based method. *Journal of Intelligent Manufacturing, 31*(5), 1243–1255.

Thacker, J. B., Carlton, D. D., Hildenbrand, Z. L., Kadjo, A. F., Schug, K. A. (2015). Chemical analysis of wastewater from unconventional drilling operations. *Water, 7*(4), 1568–1579.

Tolvaly-Roşca, F., Forgó, Z. (2015). Mixed CAD method to develop gear surfaces using the relative cutting movements and NURBS surfaces. *Procedia Technology, 19*, 20–27.

Uhlmann, E., Lypovka, P., Hochschild, L., Schröer, N. (2016). Influence of rail grinding process parameters on rail surface roughness and surface layer hardness. *Wear, 366*, 287–293.

Vryzas, Z., Mahmoud, O., Nasr-El-Din, H. A., Kelessidis, V. C. (2015, December). Development and testing of novel drilling fluids using Fe2O3 and SiO2 nanoparticles for enhanced drilling operations. In *International petroleum technology conference*. OnePetro.

Vryzas, Z., Mahmoud, O., Nasr-El-Din, H., Zaspalis, V., Kelessidis, V. C. (2016, June). Incorporation of Fe3O4 nanoparticles as drilling fluid additives for improved drilling operations. In *International Conference on Offshore Mechanics and Arctic Engineering* (Vol. 49996, p. V008T11A040). American Society of Mechanical Engineers.

Walachowicz, F., Bernsdorf, I., Papenfuss, U., Zeller, C., Graichen, A., Navrotsky, V., . . . Kiener, C. (2017). Comparative energy, resource and recycling lifecycle analysis of the industrial repair process of gas turbine burners using conventional machining and additive manufacturing. *Journal of Industrial Ecology*, *21*(S1), S203–S215.

Wang, G., Lam, H., George, A., Edwards, G. (2015, September). Performance and productivity evaluation of hybrid-threading HLS versus HDLs. In *2015 IEEE High Performance Extreme Computing Conference (HPEC)* (pp. 1–7). IEEE.

Wang, Q., Zhang, D., Tang, K., Zhang, Y. (2019). A mechanics based prediction model for tool wear and power consumption in drilling operations and its applications. *Journal of Cleaner Production*, *234*, 171–184.

Wasif, M., Iqbal, S. A., Ahmed, A., Tufail, M., Rababah, M. (2019). Optimization of simplified grinding wheel geometry for the accurate generation of end-mill cutters using the five-axis CNC grinding process. *The International Journal of Advanced Manufacturing Technology*, *105*(10), 4325–4344.

Yang, C. K., Chen, Y. H., Chuang, T. J., Shankhwar, K., Smith, S. (2020). An augmented reality-based training system with a natural user interface for manual milling operations. *Virtual Reality*, *24*(3), 527–539.

Yue, H. T., Guo, C. G., Li, Q., Zhao, L. J., Hao, G. B. (2020). Thermal error modeling of CNC milling machine tool spindle system in load machining: Based on optimal specific cutting energy. *Journal of the Brazilian Society of Mechanical Sciences and Engineering*, *42*(9), 1–12.

Zaman, P. B., Dhar, N. R. (2019). Design and evaluation of an embedded double jet nozzle for MQL delivery intending machinability improvement in turning operation. *Journal of Manufacturing Processes*, *44*, 179–196.

Zhang, G., Guo, C. (2015). Modeling of cutting force distribution on tool edge in turning process. *Procedia Manufacturing*, *1*, 454–465.

Zhang, L., Ren, C., Ji, C., Wang, Z., Chen, G. (2016). Effect of fiber orientations on surface grinding process of unidirectional C/SiC composites. *Applied Surface Science*, *366*, 424–431.

Zhang, S., Zhou, K., Ding, H., Guo, J., Liu, Q., Wang, W. (2018). Effects of grinding passes and direction on material removal behaviours in the rail grinding process. *Materials*, *11*(11), 2293.

Zhang, Y., Xu, X. (2021). Predicting the delamination factor in carbon fibre reinforced plastic composites during drilling through the Gaussian process regression. *Journal of Composite Materials*, *55*(15), 2061–2068.

Zhao, Y., Noorbakhsh, A., Koopialipoor, M., Azizi, A., Tahir, M. M. (2020). A new methodology for optimization and prediction of rate of penetration during drilling operations. *Engineering with Computers*, *36*(2), 587–595.

Zheng, F., Hua, L., Han, X., Li, B., Chen, D. (2016). Linkage model and manufacturing process of shaping non-circular gears. *Mechanism and Machine Theory*, *96*, 192–212.

Index

https://doi.org/10.1515/9783110676662-006